T0212381

Ethics on the Laboratory Floor

Ethics on the Laboratory Floor

Edited by

Simone van der Burg
IQ Healthcare, the Netherlands

Tsjalling Swierstra
Universiteit Maastricht, the Netherlands

First published 2013 by
PALGRAVE MACMILLAN

Palgrave Macmillan in the UK is an imprint of Macmillan Publishers Limited,
registered in England, company number 785998, of Houndmills, Basingstoke,
Hampshire RG21 6XS.

Palgrave Macmillan in the US is a division of St Martin's Press LLC,
175 Fifth Avenue, New York, NY 10010.

Palgrave Macmillan is the global academic imprint of the above companies
and has companies and representatives throughout the world.

Palgrave® and Macmillan® are registered trademarks in the United States,
the United Kingdom, Europe and other countries.

ISBN 978-1-349-43407-7   ISBN 978-1-137-00293-8 (eBook)
DOI 10.1057/9781137002938

This book is printed on paper suitable for recycling and made from fully
managed and sustained forest sources. Logging, pulping and manufacturing
processes are expected to conform to the environmental regulations of the
country of origin.

A catalogue record for this book is available from the British Library.

A catalog record for this book is available from the Library of Congress.

Contents

Part II Case Studies

Part III   Critical Perspectives

Figures and Tables

Figures

Tables

Introduction

Enhancing Ethical Reflection in the Laboratory: How Soft Impacts Require Tough Thinking

Simone van der Burg and Tsjalling Swierstra

The lives of contemporary men and women are entangled with technology. The alarm clock that wakes them in the morning, the clothes they put on, the food they eat, the ways they move around, the heated and air-conditioned environments they reside in, their means of communication, their entertainment (television, movies, games, books)—all are examples of technological artefacts that many people use every day. These technological artefacts don't fulfil their purposes by themselves, but depend on specific practices to do so. Sometimes these are modest, like knowing how to set the alarm; sometimes they are more complex like driving a car or handling a computer. Furthermore, all artefacts and their accompanying practices also presuppose that certain background conditions have been fulfilled (Sclove 1995). Picking up the jar of peanut butter on the breakfast table and asking where it comes from, who produced it and how, and how it arrived here reveals a long, heterogeneous and complex network of people, things and activities, such as farming skills, the production of fertilisers, flying and driving licenses, planes and cars, roads, road maintenance, petrol stations, traffic rules, police and so forth.

Looking around in our human-made environment, it is clear that we no longer live primarily in a biotope, but in a technotope. If science and technology only brought gifts, perhaps this wouldn't be a problem. But we have learned that their gifts can, and often do, have a flip side. Since the industrial revolution, wealth has grown exponentially in most parts of the world, and people have become healthier and live longer (Clark 2008). But the same revolution also brought pollution and depletion of natural resources. Technology provides easy access to the fruits of high culture to billions of people who in previous times would

1

never have had a chance to hear a concert or watch a play. But the same technology helped to create a culture of crude mass consumption. The same technological forces that enable democracy and control of the powers that be also help to monitor and control citizens. Science and technology give, but also take. They elevate, but also oppress. They create, but also destroy. As a result, the naively cornucopian conception of science and technology has gradually lost credibility: the ambiguous effects of technology call for a more discriminating attitude.

This book presents an approach to ethics and technology that initiates or enhances *timely ethical reflection together with scientists and engineers* about new and emerging technologies. Ethics on the laboratory floor is predicated on the assumption that ethical reflection during research and development can help to reduce the eventual societal costs of the technologies under construction and to increase their benefits. When ethicists are involved *during* research they can contribute to research decisions, for example with regard to what possible solutions to pursue, what research lines to prioritise or how best to design user tests. By joining these deliberations, as the contributions to this book show, ethics on the laboratory floor helps to realise the best technological alternatives.

However, the subsequent chapters also make an additional point: the range of ethical issues relevant to new and emerging technologies is broader than is usually accepted by technologists, by policy-makers *and* by most ethicists. As a result, this books also contains a plea to define ethics in a broad sense, so as not only to include obvious hazards, but also the more subtle and ambiguous changes that technologies affect in the lives of individuals and in societies. In this introduction we want to identify these types of issues, why they are often considered impossible to govern or to control, and why the laboratory is an appropriate location to confront them.

Three motives for checking scientific and technological development

There is a growing interest in tightening society's grip on the direction of scientific and technological research and development. Technology Assessment (TA) was established in the 1970s to provide anticipatory, orientational knowledge for decision-makers, and relied heavily on sociological or economic research. It promised to offer empirically informed, realistic visions of the dynamics of technological and societal change. Since then, various versions have been developed, such as awareness TA, strategic TA, participatory TA and constructive TA.

Society's interest in checking scientific and technological development was first and foremost fuelled by past crimes, mistakes and carelessness that led to technological disasters. The last decades especially witnessed a growth of activist groups, professionals and government institutions that aim to hold science and technology actors accountable for the harm they inflict. By now, they know, for example, how to make British Petroleum (BP) pay at least US$8 billion in compensation after its oil-drilling platform *Deepwater Horizon* exploded in the Gulf of Mexico in 2010. Or how to make GlaxoSmithKline (GSK) pay a US$3 billion fine for selling drugs for unapproved uses and failing to report drug safety information. Or how to prepare criminal charges against the TEPCO executives and government officials for 'professional negligence' after the 2011 Fukushima Daiichi nuclear disaster in Japan.

A second motive to increase society's influence over the scientific and technological development is economic. The days of science and technology being a toy for gentlemen of independent means are long past. Also gone are the days that society was willing to invest obediently in science and technology, trusting that somehow, sometime, this would pay off. Presently, society, whether in the form of private investors or of publicly-funded institutions, tends to expect quicker returns on investments. Private and public investors no longer write out blank cheques for science and technology, but instead make funding conditional on outcomes. As Nowotny et al. (2001) have argued, this has led to an erosion of the boundaries between science, technology, economy and society in the sense that modern scientific and technological development is increasingly propelled by the promise of future applications. Investors want to know up front how these applications will turn a profit, contribute to economic growth, help fight cancer and so forth. This requires researchers to come up with more and more specific information about the expected value of their findings and products.

A third, and more recent, motive to extend control over science and technology is the growing public awareness of the manifold ways technologies help to shape the way modern humans live their lives. Technologies not only produce economic value and hazards, they also affect the ways in which we relate to the world, to our fellow human beings, and even to our own bodies and inner worlds. Research in Science and Technology Studies has provided rich descriptions of how technologies are value-laden and actively influence how we live, how we act, how we relate to each other, how we understand the world and our place in it, and what we aspire to, desire and hope for (Akrich 1992; Harré 2002; Keulartz et al. 2004; Latour 1999; Verbeek 2005, 2011). Think about how the spread of portable communication devices (smartphones, laptop computers) has helped to blur the boundary between work and free time, between public and private domains. How Facebook has modified traditional understandings of 'friendship' and 'privacy'. How telecare technologies are changing current care practices. How reproductive technologies have changed the experience of having children to 'choosing or deciding to have' children, with maybe the 'designer baby' as a next step. What these examples show is that technologies influence who we are and how we arrange our lives. With technology becoming all-pervasive, the need grows to reflect on these broader changes as well, and to take them, to some extent, into our own hands.

Soft impacts and responsible innovation

Three motives have been listed for society to try to check and influence the development of science and technology. But society has not been equally successful on all accounts. As the examples of BP, GSK and TEPCO illustrate, society seems to have learned some lessons with regard to technological hazards, and is now improving its self-defence techniques and its retribution mechanisms to deal with those cases where technologies inflict harm. It is also managing to press its 'valorisation agenda' onto science and technology, although here it is still struggling to find the adequate tools. The real problem, however, seems to be with the broader changes that technology brings about in human lives. Although many citizens seem particularly concerned about these, they are hardly taken up by technology actors or policy-makers. Why is that?

Extending society's influence so as to deal with how technology affects how human beings live, faces unexpected theoretical and practical challenges. Compared to clear-cut technological hazards and economic costs/benefits, technologically-induced changes in our way of perceiving, acting, valuing, hoping and relating, etc. are much harder to identify, to acknowledge, to evaluate and to anticipate. In addition, it is much harder to allocate responsibility for them. The reason for this is that both technology actors and policy-makers tend to deny any responsibility for these types of consequences of the technology—leaving it to the technology users to deal with technology's cultural impacts. In other words, they tend to treat these types of impacts as private, rather than public, concerns. Why? There seem to be at least three reasons for their unwillingness to share responsibility for cultural impacts (Swierstra and Te Molder 2012).

The first reason relates to the acceptance of *values*. At least in developed countries, technology actors are, by now, accepting responsibility for avoiding or minimising risks to our environment, health and safety, and society is willing to hold them accountable when they fail to live up to their responsibilities. What the examples of BP, GSK and TEPCO share is that all stakeholders assessed the consequences to be undesirable because the values involved—health, safety and sustainability—are non-controversial and widely shared. This agreement seems to be a precondition for technology actors to accept (some) responsibility and for government actors to hold them accountable. By contrast, when we consider the changes that technologies bring about in human lives, for example in the ways we have children or start relationships, the implied values are contested, conflicting or changing. With respect to such ambiguous consequences, it is relatively easy to deny responsibility because it is unclear whether harm has, indeed, occurred. Is it better or worse to choose to have children rather than having them? We now frown on the idea of human enhancement, but maybe tomorrow this will be a moral requirement for parents?

A second reason why these ambiguous consequences of science and technology are often dismissed as being private is that they are usually difficult or impossible to quantify. Both technology actors and policy-makers prefer impacts to be quantifiable: there is x chance that y deaths will occur. Risk assessments give a fairly precise estimation of how unhealthy, how unsafe and how destructive to the environment a technology will be, and how large the chance is that these hazards will occur. Of course, the data necessary to do risk calculations are not always available, in which case we don't speak about risk, but about uncertain or unknown hazards. But, even then, the assumption is that these hazards are 'in principle' calculable, if only we could have that information.

This is much less obvious in the case of the manifold ways in which our motivations, habits, (emotional) experiences, ways to act or interact, are affected by technologies, such as the pervasive and inescapable influence of mobile phones, or the way conceptions of friendship and intimacy change under the influence of social media like Facebook and Twitter. In the case of these broader impacts qualitative rather than quantitative discourse seems more apt. Unfortunately, qualitative discourse lacks the mark of objectivity that quantitative discourse has. This lack of objectivity makes it hard for these kinds of impacts to gain access to the public agenda, or, to be more precise, to be accepted as a matter of concern by technology actors and policy-makers.

Third, for hazards to be taken seriously as public concerns it seems necessary that they are 'determined' by the technology. In discussions about the responsibility of technology actors and policy-makers, an instrumentalist vision on technology still prevails. This outlook on technology is caught perfectly by the US bumper sticker of National Rifle Association supporters: 'Guns don't kill people; people kill people' (Swierstra 2005, pp. 18–28). According to this instrumental perspective on technology, if something goes awry the user is to be blamed not the maker or provider of the instrument. Unfortunately, the broad ways in which technology changes human lives are never determined solely by technology; they are produced by an interaction between science, technology and human agents. For instance, artefacts can constrain, enable, stimulate, invite or seduce their users to perceive the world, or interact with it in specific ways. Low-fat crisps seduce the buyer to eat more; cheaper technologies stimulate their increased use; the television changes how politicians are perceived; and so forth. In all of these examples human agents do something, so they are responsible; yet, they would have acted differently if the technology was not available. Consequences can therefore not be tied unequivocally to scientists and developers of technology, or to policy actors. These are subsequently all too glad to shift the burden completely to the users thus, again, privatising the issues.

In sum, to be taken seriously by technology actors and policy-makers, and for them to accept responsibility, technology's consequences should conform to at least three implicit criteria: they should involve indisputable instances of harm, they should be quantifiable, and a clear causal link between technology and consequences should exist. Only then do these impacts seem 'hard' enough to warrant public concern; if, however, these impacts are morally ambiguous, qualitative, and co-produced by artefact and user, then they are dismissed as too 'soft' and delegated to the private sphere.

In this book, however, we *are* interested primarily in these latter types of effects. Their alleged 'softness' does not diminish their importance. As they refer to changes that technologies affect in the ways in which people lead their lives, we believe they deserve more public attention. Unfortunately, most forms of TA remain geared to technology's 'hard impacts', which, by definition, require little ethical reflection. The broader ways in which technology affects human life—which we will call here 'soft impacts'—concern topics as fuzzy, subjective and private as the good life, and they rarely come into play. However, this may be changing. Increased attention to soft impacts can be recognised in the recent transition from TA to so-called 'responsible research and innovation' (Von Schomberg, in press) Responsible innovation refers to efforts that try explicitly to construe mechanisms for making people responsible for 'hard' impacts, but also for the 'soft' impacts.

Ethics on the laboratory floor fits squarely with the general ambition of responsible innovation, for it aims to use moral philosophy to distinguish issues relating to a developing technology that demand ethical reflection, and to provide tools to enhance reflection about them among scientists and engineers. This is a precondition for making responsible choices, or creating responsible practices of research and development.

The good life

The contributions to this book focus on the 'soft impacts' of technology and help to show why they deserve attention. While it is difficult to put them on a political agenda, they have to be dealt with somehow. This book proposes that they can be handled on the laboratory floor. Even if it is hard to make policy-makers think about soft impacts, the authors in this book take it that scientists are able to recognise their importance and are willing to take them into account in the further planning of their research.

A laboratory ethics therewith abandons the role of the ethicist as an outsider and exchanges it from someone on the fringe to an insider, a collaborator, a colleague in an interdisciplinary set-up. In the parameters of the traditional division of moral labour, scientists are left to their own devices as long as they restrict themselves to the facts; value issues are then delegated to the ethicist, policy-maker or politician. But with an acceptance of the view that scientific and technological research and development are never value-free, the role of ethicists will also change. It can no longer be the role of the ethicist to import norms and values from the outside. Instead, her/his role is to *engage with the researchers on the laboratory floor,* and to help articulate and discuss existing moralities, as well as the possibilities for the future. The ethicist enters into an ethical dialogue with scientists and technologists in which relevant morals are articulated, discussed, maybe changed, and interpreted. In this sense, she/he participates in research and then also has to accept co-responsibility for the outcomes.

The chapters in this book discuss ambiguous and contested consequences of new technologies (the soft impacts). It is probably the most important message of this book that these deserve attention, and several methodologies are presented for distinguishing them. These soft impacts, in our view, do not fit into just any approach to ethics. The orientation on harm (hard impacts) is most compatible with consequentialism, as this way of framing problems is very commensurable with the preference for quantifiability that we observed before. A deontological approach to ethics, however, would ensure that the rights of important stakeholders are secured, which is particularly commensurable with a legalistic approach, based on rules that define a negative freedom. It presupposes that if you keep to the rules, for the rest you are free to do as you choose. What is not forbidden as a transgression of other stakeholders' rights, or which does not harm other people, falls outside the scope of morality and ethical discourse.

In the basic grammar of these two ways of doing ethics, we also see reflected the basic features of a modern, liberal, free market society: a calculating approach bent on maximising benefits, only restricted by a limited set of basic rules that allow fair play and competition. Soft impacts, however, are hardly discussed in deontological or consequentialist perspectives to ethics. These 'soft impacts' fit into a psychological approach to morality, which are represented by forms of good life ethics. The term 'soft impacts' refers to effects that concern human capacities such as experience, motivation, emotion, habit, perception, ways of relating to others and deliberation about action. These capacities play a central role in human moral psychology and therefore—thinking along a virtue ethical line—can be expected to influence how people understand the good life, and whether and how they are able to realise it in a world with technology.

While deontological and consequentialist approaches to ethics focus on individual actions, a good life ethics will consider decisions against a background of experiences, emotions, relations and beliefs, which are formed and cultivated in a social environment. In the specific context in which an individual is located, certain actions are revealed as possibilities and others as difficult or impossible to realise. This background and context are—to a certain extent—constitutive of the decisions that individuals take, even though they do not determine them. While attempts to reflect on technology from a good life perspective to ethics are rare, we think that it leaves room to reflect on technologies that are present in a context and how they steer what individuals want, aim for or perceive to be good, as well as how they form the social context in which people associate with one another. A good life ethics also offers tools to reflect on these contexts and the (im)possibilities that they create for individuals and societies to realise a life that they find worthwhile.

All of the chapters bring forward considerations about soft impacts. While authors seldom place their considerations about values in a specific ethical discourse, we, as editors, want to suggest to think about them in the perspective of a good life ethics. In a good life ethics a moral vocabulary is developed that is sufficiently broad to accommodate considerations about soft impacts. Moral statements take the form of advice or recommendations, and reveal what is good and desirable to strive for. This positive heuristic fits inside a laboratory where scientists are working on a technology that embodies promises for the future, and which should be directed in better, rather than worse, ways. It is therefore a specific type of ethics that would fit the content of the issues that are at stake, as well as the context in which they are to be discussed.

The location of the laboratory

There are, of course, many moments in the development of a new technology when it could be useful to enhance ethical reflection. But we think the laboratory is a particularly good place to do it. During research, a technology has not yet finished developing: it is still relatively easy, and cheap, to bow the further development of the technology in another direction. If such a change in the development trajectory of a new technology is motivated by a joint reflection of scientists and moral philosophers, scientists may not experience it as humiliating criticism, but as constructive feedback, which makes them feel more satisfied with their own work. If scientists and engineers are convinced of the underlying reasons and understand how their actions could contribute to making a better technology, it is easier to integrate ethical reflection into their work.

Ethical reflection is especially helpful when it concerns concrete research decisions or material aspects of the technology for which scientists or engineers already feel responsible. We think moral philosophers have a chance to come up with such topics for reflection if they look inside the laboratory. In a laboratory it is possible to see how the technology works in the experimental set-up, which allows the values materialised by the technology to be discerned, which may be important to take into account during future stages of research. Furthermore, scientists discuss their laboratory experiments in meetings and there ethicists can hear how scientists think about their research, what its challenges are and what are considered 'good' ways to meet them. Distinguishing these values and enhancing reflection about them helps to make scientific engineers more aware of the societal relevance of their research activities, which adds another dimension to the ways in which they normally think about what they do. Such reflections that relate directly to the challenges that they face, the actions they are engaged in and the choices they make about future research steps have a good chance of becoming included in their self-evident understanding of their responsibility.

There are also obstacles to the work of moral philosophers in the laboratory. Moral philosophers may be interested in broadening the future orientations in which scientists already engage in order to include thoughts about how the technology to which they contribute could change society or individual human life. But research is organised in carefully designed stages, which discourages scientists from injecting such future-oriented ethical reflection into the research agenda. Invitations to think about ethical issues are therefore frequently declined and postponed to a later stage of research, or responsibility for them is ascribed to other people. It demands a reorganisation of the research protocol in order for scientists to accept ethical reflection as something that is their business to engage in.

Laboratory work, furthermore, is often carried out by many hands, including technicians, and junior and senior researchers (and full professors), as well as non-scientific members of the research consortium, such as representatives of research funding institutions, members of the industry and—sometimes—future users. This makes it hard to make anyone responsible for the issues that demand ethical reflection: whose responsibility is it to think about these issues and to actually take care of them in the course of shaping the technology? The problem for the moral philosopher is to find a vocabulary that makes it possible for members of such a complex research consortium to accept and appropriate a moral issue as an issue that they should think about as a collective. The term 'responsibility' still has quite an individual meaning, which makes it hard to grasp in a context where responsibility is usually shared.

A third obstacle to the acceptance of responsibility for the ethical issues is that research consortia are often characterised by hierarchical relationships. The researcher who carries out the laboratory experiments is not the one who formed the research project, or who takes the final decisions about the research direction to pursue, or the applications that are to be realised. The challenge for the moral philosopher is to involve the right people in a reflection about the values inherent to the research activities and the technology to which they contribute.

These obstacles are, however, not only obstacles for moral philosophers, but also for scientists themselves. Scientists have difficulty to respond to the societal demand for responsible innovation, which requires being reflective—and take considered and well-informed decisions—about what research is worth pursuing, to what applications it should contribute and how these applications should be formed. Philosophers who are engaged in the laboratory may assist scientists and engineers in responding to this societal demand by becoming aware of the obstacles and trying to develop a more precise insight into the potential abilities of their research to change society, as well as individual human life, and develop a broader approach to their responsibility. The exchange with moral philosophers may, therefore, be experienced as helpful in responding to societal demands and they can offer new food for thought, which enhances creativity.

The contributions to this book

The contributions to this book discuss what ethics on the laboratory floor is, what it aims to do and what the role of an ethicist in this location can be, but also what the limitations of an ethics on the laboratory floor are. The authors are scholars with backgrounds in diverse disciplines, including philosophy of technology, science and technology studies, technology assessment, political science and history of science. During the past few years we have regularly engaged in debates on the topic of this book with all of them during workshops and conference meetings, as well as during more informal gatherings. With some of them we formed a research group called Ethics and Politics of Emerging Technology (EPET) that meets several times a year and which held its first international conference in July 2012. Because of this continued interaction and exchange, the chapters of this book reveal a coherent whole, while also leaving space for disagreement and debate.

The book is ordered into three parts. In the first part we offer four perspectives on the role that moral philosophers should take in the laboratory. Bernadette Bensaude-Vincent contrasts ethics on the laboratory floor with ethical, legal and social issues/aspects (ELSI/ELSA) research, which aims to accompany research into a new technology with research into the ethical, legal and social issues (or aspects) relating to it. Research into these 'issues' that are supposed to materialise if the technology is realised, according to Bensaude-Vincent, is often based on too speculative assumptions about the future. Instead, she proposes that ethics on the laboratory floor focuses on present scientific practices. A laboratory ethics is conceived in this contribution as an attempt to break open the social world of the laboratory and enhance reflection about visions of the future that guide the actual actions of scientists.

In Chapter 2 Armin Grunwald draws attention to the ethical responsibility of scientists. Scientists and engineers have an ethical responsibility, especially because the technologies that they make may bring about situations that conflict with the generally accepted morality in a society. Moral philosophers who are engaged in the laboratory have, according to Grunwald, the function of 'tracker dogs' who may warn society about the conflicts and dilemmas that may be raised by a new technology, and who may identify, clarify and 'treat' those conflicts. In Grunwald's perspective, the moral philosopher who is engaged in the laboratory primarily deals with the values of people—scientists, engineers, (future) users, politicians, society at large—and the ways in which they talk about what they value in a technology.

Peter-Paul Verbeek, by contrast, argues that technology itself can materialise morality. Taking distance from the metaphysical subject–object distinction, which framed the separation between morality and technology, Verbeek argues that human beings, as well as technologies, become what they are in interaction with one another. Rather than focusing solely on the people and the moralities that they express and hold dear, Verbeek thinks it is distinctive for the role of moral philosophers that they focus on the technologies that are being created and the values that they embody. Verbeek's examples of how technologies materialise morality are convincing because they are relatively close to their implementation, and it is possible to distinguish quite precisely how they change what choices people think they have to make and how they make them. Marianne Boenink, however, thinks the uncertainty of technologies during research is their most distinctive characteristic. She makes it a distinctive task of moral philosophers to specify what it is exactly that scientists and engineers are creating, based on their (often diverse) ways of speaking about it, as well as what can be seen in the laboratory.

The approach that Boenink advocates is practice-based in the sense that she understands science and engineering as a practice in which people interact, and whose interaction might change because of the introduction of a new person: the ethicist. The design of her chapter is also practice-based, for she derives the distinctive family resemblances of the perspective that she advocates from actual 'examples' of ethics on the laboratory floor work. On the basis of her selection of examples, Boenink distinguishes, in addition to specifying, five tasks that characterise—but do not exhaust—the role of ethicists at this location: reconstructing, probing, broadening and converging/aligning.

The second part of this book provides case studies that offer concrete examples of the work of moral philosophers on the laboratory floor. However, because the authors of this book began their studies before methodologies were available for a laboratory ethics, these case studies do not provide simple applications of the methodologies provided in section one. It is for this reason that the case studies do not apply theory to a case; rather, out of the case descriptions arise proposals as to how moral philosophers on the laboratory floor could carry out their job.

The contribution by Simone van der Burg, for example, discusses ethics in laboratory research conducted at the University of Twente (the Netherlands) into an acousto-optic device for the non-invasive monitoring of chemical substances in blood, such as oxygen and (in the future) glucose, cholesterol or lactate. The methods she uses suit the description of 'specification', 'probing' and 'broadening', which Boenink distinguished. But she also adds something extra: she explores the past of this technology. Acousto-optics is related to a family of optical technologies, some of which have already developed into devices that have been used on patients for three decades. This 'ancestry' is informative about the values it may express when it develops into a usable product, and which it will—in Peter-Paul Verbeek's words—'materialise'. This concentration on the value materialised in technological artefacts fits with Verbeek's interest in the morality of things, but approaches it in quite a different way, which may be particularly useful in a research phase.

Xavier Guchet, by contrast, focuses not on the past of technologies, but is interested in visions of the future that guide present research and development in nanotechnology, especially in the artificial synthesis of the flagellar motor of bacteria by LAAS, a well-known French laboratory. His contribution elaborates Bernadette Bensaude-Vincent's pragmatist approach to a laboratory ethics, which avoids speculation about the future, but concentrates on the ways in which research takes shape in current evaluative discourses in which the dichotomy between the natural and the artificial play an important role. It invites scientists to reflect on the values that underlie their research on the basis of a reading of philosophical texts—for example, by Hannah Arendt—about the nature–artefact distinction, and tries to come to grips with the evaluations that guide their work and introduce broader societal perspectives into those evaluations.

Chapter 7, by Fern Wickson, describes the ethical work of social scientists engaged in ecotoxicology research in a Norwegian laboratory who aim to create opportunities for the emergence of a more ethically responsible ecological governance of technology development. To this end conflicts are explored between the use of waterfleas in ecotoxicology experiments and environmental ethics standards. As a co-worker at the laboratory bench Wickson invites scientists to reflect on whether the environment can be used by human beings, or if the organisms and systems involved have a value independent of human aims and desires. Her identification and clarification of ethical issues, on the basis of already well-known controversial debates about animal welfare within environmental ethics, suits the role that Grunwald sketches for ethicists on the laboratory floor. Yet, her efforts to become part of the research team and to enhance ethical reflection as a co-worker in the laboratory fits better in the practice-based approach that Marianne Boenink describes.

Federica Lucivero presents an extensive case study on an emerging biomedical technology for early diagnostics and health monitoring called Immunosignaturing (ImSg), which is being investigated at Arizona State University, USA. Her chapter describes an example of a practice-based approach to laboratory ethics, which aims for a production of technology that is more aware of—and reflective about—how it will shape society and individual life. Her case study criticises an ELSI approach as being too speculative about the future. Rather than focusing on the generalised and simplified expectations that are communicated to the public, Lucivero articulates the more diverse and more complicated expectations that scientists articulate within the laboratory. This case study shows how the acts of specifying (the possible futures of the technology) and broadening (of the reflection about those futures) could take shape in the work of a moral philosopher in the laboratory.

In Chapter 9 Lotte Krabbenborg presents a case study on the development of the lithium chip. The idea behind the lithium chip is to create point-of-care testing for people who suffer from a bipolar disorder and use lithium as medicine. It would allow patients to test themselves at any place—at home or in a professional setting—without the interference of a (physical) laboratory. Krabbenborg organised workshops to discuss the ethical issues relating to this technology, which involved diverse stakeholders. To shape these workshops she used John Dewey's notion of a 'dramatic rehearsal', which refers to an ideal mode of deliberation required when people find themselves in situations in which it is not clear how to act, what to value or which ends to pursue. Krabbenborg's chapter illustrates what the act of 'broadening' could involve, which many ethicists on the laboratory floor take to be useful in enhancing ethical reflection.

The third part contains critical perspectives. It discusses limitations of an ethics on the laboratory floor. Arie Rip points out in Chapter 10 how normativity is an anthropological category, which means that it is pervasive. Ethics only articulates and explores limited parts of that broad normativity, leaving other parts to disciplines like law and sociology. As a result, he argues, ethics turns a blind eye to forms of normativity that are, indeed, relevant for dealing with emerging technologies, and he discusses two examples of such ignored articulations of normativity, one in relation to the promises that are made on behalf of emerging technologies and the other in relation to the activities and roles of agents of constructive TA.

In the final chapter, Alfred Nordmann is concerned that ethics on the laboratory floor will function as an excuse to stop considering science and technology policy. Most choices that concern responsible innovation have already been made before research is done in the laboratory. While laboratory work could also benefit from reflexivity, a broad responsibility for the world and the lives that are constituted through design processes needs to address a broader group of actors, including science policy-makers, market researchers, advocacy groups, and developers of devices (but also of packaging and advertising), and also including early adopters and other buyers who establish the patterns of use that settle the definition of the designed artefact. Ethics on the laboratory floor, according to Nordmann, addresses too limited issues and addresses just a small selection of the actors involved.

Acknowledgements

We thank the Netherlands Organisation for Scientific Research for funding research of the editors that lead to this volume, and the Netherlands Graduate Research School of Science, Technology and Human Culture (WTMC) and 3TU Center for Ethics and Technology for funding the seminar that sparked this volume. We also thank Audrey Rhodes for her improvements of the quality of English of the contributions to this book.

Further reading

Beck, U. (1992) *Risk Society. Towards a New Modernity* (Cambridge: Polity Press).

Grin, J. and Grunwald A. (2000) *Vision Assessment: Shaping Technology in the 21st Century Towards a Repertoire for Technology Assessment* (Berlin: Springer Verlag).

Lucivero, F., Swierstra, T. and Boenink, M. (2011) 'Assessing Expectations: Towards a Toolbox for an Ethics of Emerging Technologies', *NanoEthics*, 5(2): 129–41

Pols, J. (2012) *Care at a Distance. On the Closeness of Technology* (Amsterdam: Amsterdam University Press).

Turkle, S. (2010) *Alone Together. Why We Expect More from Technology and Less From Each Other* (New York: Basic Books).

Wynne, B. (1996) 'Misunderstood Misunderstandings. Social Identities and Public Uptake of Science', in Irwin, A. and Wynne, B. (eds) *Misunderstanding Science? The Public Reconstruction of Science and Technology* (Cambridge: Cambridge University Press).

Wynne, B. (2001) 'Creating Public Alienation: Expert Cultures of Risk and Ethics on GMOs', *Science as Culture*, 10(4): 446–81.

Wynne, B. (2006) 'Public Engagement as a Means of Restoring Public Trust in Science—Hitting the Note, but Missing the Music?', *Community Genetics*, 9(3): 211–20.

References

Akrich, M. (1992) 'The De-Scription of Technical Objects', in Bijker, W. and Law, J. (eds) *Shaping Technology/Building Society: Studies in Sociotechnical Change* (Cambridge, MA: MIT Press).

Clark, G. (2008) *A Farewell to Alms: A Brief Economic History of the World*. (Princeton, NJ: Princeton University Press).

Harré, R. (2002) 'Material Objects in Social Worlds', *Theory Culture Society*, 19(23): 23–33.

Holbrook, J. B. (2005) 'Assessing the Science–Society Relation: The Case of the US National Science Foundation's Second Merit Review Criterion', *Technology in Society*, 27(4): 437–51.

Keulartz, J., Schermer, M., Korthals, M. and Swierstra, T. (2004) 'Ethics in a Technological Culture. A Programmatic Proposal for a Pragmatist Approach', *Science, Technology and Human Values*, 29(1): 3–29.

Latour, B. (1999) *Pandora's Hope. Essays on the Reality of Science Studies*. (Boston, MA: Harvard University Press).

Nowotny, H., Scott, P. and Gibbons, M. (2001) *Rethinking Science. Knowledge and the Public in an Age of Uncertainty* (Cambridge: Polity).

Sclove, R. E. (1995) *Democracy and Technology* (New York: The Guilford Press).

Swierstra, T. (2005) 'Hoe samen te leven met techniek?', *Wijsgerig Perspectief*, 45(3): 18–28.

Swierstra, T. and Te Molder, H. (2012) 'Risk and Soft Impacts', in Roeser, S., Hillerbrand, R., Peterson, M., and Sandin, P. (eds) *Handbook of Risk Theory. Epistemology, Decision, Theory, Ethics and Social Implications of Risk* (Dordrecht: Springer).

Von Schomberg, R. (in press) 'Prospects for Technology Assessment in a Framework of Responsible Research and Innovation', in Dusseldorp, M. and Beecroft, R. (eds) *Technikfolgen abschätzen lehren: Bildungspotenziale transdisziplinärer Methoden* (Wiesbaden: Vs Verlag).

Verbeek, P. P. (2005) *What Things Do. Philosophical Reflections on Technology, Agency and Design* (University Park, PA: Pennsylvania State University Press).

Verbeek, P. P. (2011) *Moralizing Technology* (Chicago: University of Chicago Press).

Part I

Moral Philosophers
in the Laboratory

1
Which Focus for an Ethics in Nanotechnology Laboratories?

Bernadette Bensaude-Vincent

Introduction

Over the last decade, ethics has been institutionalised as an integral part of nanotechnology research. This integration was a response to the alert 'Mind the gap', which was launched in 2003. 'As the science leaps ahead, the ethics lags behind. There is danger of derailing NT [nanotechnology] if the study of ethical, legal, and social implications does not catch up with the speed of scientific development' (Mnyusiwalla et al. 2003). One year later, the report of the Royal Society and Royal Academy of Engineering in Great Britain *Nanoscience and Nanotechnology: Opportunities and Uncertainties* also pointed to the gap between technology and ethics, and made a similar claim about the gap between science and the public (Royal Society and Royal Academy of Engineering 2004).

The implication of ethics in the regime of 'normal science' has been favoured by the repeated calls for 'responsible innovation' from the promoters of nanotechnology initiatives. Such calls testified, to a certain degree, to the awareness that previous practices of innovation 'in the name of progress' were not responsible enough. It is clear that the motto of the 1933 Chicago World Fair—*science finds, industry applies, man conforms*—is no longer acceptable. However, over the last few decades the ethical concern about scientific and technological choices has prompted new research directions. Instead of the creation of ethics committees, concern around responsibility resulted in the implementation of research programmes on ethical, legal and social implications (ELSI) being integrated in the national nano-initiatives. Thanks to the generous funding of the nano-initiatives dedicated to ELSI, nanoethics

21

has become a booming research field with a growing community of scholars and a journal, *NanoEthics*, which was founded in 2007.

As they deal with effects, ELSI research programmes are focused mainly on the technological applications of current research. They tend to forget the present in favour of speculative futures. To what extent could the proposal of an ethics on the laboratory floor help to refocus on the present and foster ethical judgements? What kind of ethics could be developed on this site?

In this chapter I will first argue that, when taken seriously, the proposal of a laboratory floor ethics may help to move beyond the limits of ELSI programmes. However, it would be misleading to view the laboratory as the cradle or birthplace of technological applications that will shape the future of humankind. Rather, they mirror the present view of the future. I will argue that an epistemological analysis of technoscientific activity is a helpful detour to refocus ethics on the design of objects. Nano-objects will be redefined as carriers of meanings and valuations that need to be made explicit, then articulated and submitted to a moral evaluation.

The ethical turn in science and technology

ELSI research broke with the tradition of applied ethics. Instead of looking at the arsenal of ethical doctrines—ethics of virtues, principlism or consequentialism—in search of the best candidate to apply to the case of nanotechnology, ELSI programmes favour a bottom-up approach. Taking the agenda of nano-initiatives into account, the task is to identify and anticipate the impacts of nanotechnology on society at large. As pointed out by Alfred Nordmann, the ELSI approach also broke with the debates raised around the methods of technology assessment developed in the 1980s, concerning the right moment for a technology to be subjected to social control. In ELSI, the right moment is upstream, before the technology is disseminated or even proved feasible through anticipation of its potential consequences. It is described as a 'proactive' rather than merely a reactive attitude for promoting responsible innovation and it is, in fact, shaping the future (Nordmann 2010). Although this approach can be seen as a consequentialist approach to ethics, in practice it is essentially a prospective exercise on the anticipation of potential consequences, risks or conflicts to the values of society as it is now. This prospective approach, already experimented for the development of genomics, calls for social scientists, philosophers, lawyers and economists. Ethics is no longer solely in the hands of professional ethicists. Social scientists have been embarked on the board of research laboratories, not just to observe scientists in action, but also to engage in dialogue and try to trigger reflections about the effects of their research on society. The mission of ELSI researchers is to identify and clearly artic-ulate a number of major issues raised by the development and diffusion of the emerging technology.

The result is the establishment of a list of problems—toxicity and environmental effects, safety and security, privacy, human enhance-ment, intellectual property and global justice. The list has been adopted and used in national and international reports or public debates on nanotechnology with few nuances. It quickly became a standard check-list, such as the one we use to travel safely. It proved crucial in raising concern among policy-makers, scientists and industrialists, and in engaging them in public debates. To a certain extent it kicked off an attempt towards the co-construction of science, technology and society (European Union High-Level Expert Group, 2004).

The limits of the ELSI approach

The checklist, however, is no more than an early diagnosis of poten-tial upcoming issues. It has nothing to do with normative ethics. A list of problems does not provide clues to their solutions or guidelines for action. And there is a great distance between the realisation of the poten-tial risks raised by nanotechnology and the implementation of legal measures to prevent them. First, ELSI has to be completed by regula-tions, codes of conduct and guides of good practices, all of which rely on ethics principles and values applied to the field of nanotechnology. More than an ethical guide, the establishment of a checklist of potential issues is a useful tool for risk governance. It looks like a routine prospec-tive exercise in industry or business, and it belongs to the methods of management more than to ethics.

Second, ELSI programmes, as they have been conducted over the last decade, raise inner tensions among the social scientists engaged in them. These programmes were set up in order to address all soci-etal, ethical and legal issues upstream during the earliest stages of research. However, the issues that have been identified and discussed in most publications and conferences are only concerned with appli-cations of nanotechnology research that anticipate the potential future undesirable consequences of their diffusion in society. With this focus on future applications, ELSI programmes relied on the timeline of the nano-initiative roadmap with its four steps: (i) passive nanostructure;

(ii) active nanostructures; (iii) systems of nanosystems; and (iv) molecular systems with emerging functions (Roco et al. 2000). Many ethical reflections and debates thus concentrated on nanorobots and enhanced human capacities, and became more and more prospective and speculative. Fascinating futures of autonomous robots, post-humans and immortal life have attracted a lot of public and scholarly attention over the last decade.

Thus, ironically, while ELSI programmes meant to anticipate the potential adverse effects of nanotechnology, they resulted in endorsing and validating incredible futuristic suppositions. As Alfred Nordmann argued, they generated a perverse effect as they transformed the conditional tense following 'if' into a present tense describing the current state of affairs:

> The true and perfectly legitimate conditional 'if we ever were in the position to conquer the natural ageing process and become immortal, then we would face the question whether withholding immortality is tantamount to murder' becomes foreshortened to 'if you call into question that biomedical research can bring about immortality within some relevant period of time, you are complicit with murder' – no matter how remote the possibility that such research might succeed, we are morally obliged to support it. (Nordmann 2007; Rip 2009)

This spectacular reversal of the ethical question about research aimed at human enhancement relies on a pre-emption of the future, a potential and improbable future being posed as real. In their prospective effort, ELSI research programmes, in general, tend to assume that our technological future is predetermined and that society has to regulate or modulate its development. This deterministic assumption underlies the management of the consequences of nanotechnology applications. Although it provides a sense of control it boils down to generating social acceptability. Moreover, this sense of control is partially blind to past experiences, which taught us that the future is not always in our hands, and that historical contingencies and complex interactions deeply affect the course of technological changes. Thus, ironically, the concern with the future is divorced from the present, as well as the past.

Third, the focus on futuristic applications of nanotechnology generated an additional perverse effect. Although ELSI research was launched to address issues upstream in a proactive manner, it failed to interest the actors upstream. Prospective exercises do not appeal to researchers at the bench. Laboratory workers tend to think that they are not concerned with issues raised in ELSI research, as long as they are not working on futuristic applications, such as nanorobots or brain implants that will enhance human performances, etc. (Sauvage 2008). They tend to consider ethicists as professional experts in abstract speculations and often call for a division of labour: we are dealing with technical issues, and it is the ethicists' and politicians' responsibility to deal with moral and societal issues. Or, when they have been convinced that they should engage in the ethical reflection on their own research agenda, they often consider ethics as a problem-solving activity based on the model of scientific activity. And they expect that an ethics toolkit would help them behave as 'responsible scientists'.[1]

Finally, ELSI programmes are also unsatisfactory for professional ethicists. In most sectors of applied ethics—medical ethics, environmental ethics, neuroethics, etc.—ethicists have the choice to develop a variety of moral theories, such as principlism, consequentialism, ethics of good life or ethics of care. In ELSI programmes ethics seems to be concerned exclusively with consequentialism. More precisely, the anticipatory attitude relies on a managerial translation of consequentialism understood as the balance between the moral acceptability of risks and the desirability of benefits. Ethical deliberation boils down to cost–benefit analysis, a ritual exercise in the conduct of business affairs. Ironically, if ELSI programmes are of interest to professional ethicists it is because the two aforementioned 'perverse effects' raise a classical issue of moral philosophy: the issue of unintended effects of actions linked with the notion of 'moral luck', which begs for a moral judgement (Williams 1981).

Why ethics on the laboratory floor?

If ELSI programmes are unsatisfactory both for ethicists and for scientists because they are too future-oriented, do we have a better chance of success by focusing on the present, on the actual research presently conducted on the laboratory floor?

In order to raise ethical issues on the laboratory floor, one has to dismiss the traditional linear model with which the process of technology development is frequently represented. This model has its conventional sequence: basic science – applied science – prototype – industrial development – commercial applications. As long as the linear model prevailed there were no reasons why ethicists should play a role in the laboratory. The traditional ethos of science, as defined by Robert Merton

and based on the experience of one century of academic research, suf-
ficed to cope with most ethical concerns (Merton 1986). According to
Merton, basic science is supposed to be 'pure', neutral, amoral, while
applications would be good or bad depending on their uses and on the
context of use.

If it is taken seriously, the proposition to bring ethicists onto the
laboratory floor requires the linear model to be brought into question
together with the 'purification' of science, which it implies. Introducing ethics onto the laboratory floor presupposes that laboratory research
is not shaped exclusively by epistemic values, such as truth, empirical
adequacy and objectivity, or simplicity, generality and accuracy (Carrier
et al. 2008). Research agendas in nanotechnology are also ruled by social
values and driven by great expectations of medical, environmental or
practical applications. Research in nanotechnology is by no means a
disinterested pursuit of knowledge. Does this mean that researchers on
the laboratory floor are committed to non-epistemic values? And to
what extent may the participation of non-epistemic values raise ethical
concerns?

It is clear that the regime of research, which prevails in
nanotechnology—with interdisciplinary teams and large networks,
start-ups and industrial contracts—seriously challenges the norms that
Robert Merton ascribed to science, such as communalism, universalism,
disinterestedness and organised scepticism. When researchers are deeply
entangled in industrial ventures, encouraged to patent their results or to
keep them secret, other norms are emerging, such as reliability, accountability or reflexivity, which are not specific to the scientific profession
(Ziman 1994, 1996).

However, the changing ethos of scientists and their new professional identity is not just a consequence of the practical orientations
of research in nanotechnology. In fact, nanotechnology, like a number
of technoscientific research areas, combines science and technology in
many different ways. Their interplay shapes research agendas, changes
the criteria of success and also implies a new definition of what matters
in research—performing more than understanding. It also determines
what counts as a proof and it brings about a new relation to research
objects.

A few words about technoscience in general may be a useful detour
to shape the project of ethics on the laboratory floor. Although no laboratory in the world uses the term 'technoscience' in its title because
it has depreciative connotations, this term can be used as a descriptive

category to characterise the epistemology of research in a number of areas, such as genomics, nanotechnology or synthetic biology.[2]

The term technoscience was coined in the 1980s by the Belgian philosopher Gilbert Hottois to refer to science that is done in a techno-logical setting with high-tech or mathematical instruments, and which is driven by the perspective of practical applications (Hottois 1984). Bruno Latour made extensive use of the term 'technoscience' in his effort to fight against the myth of pure science. All sciences and all technologies are hybrids, as science, technology and politics cannot be separated from one another (Latour 1987). The term 'technoscience' does not just mean that science and technology cannot be held apart and that the distinction between pure and applied research is a myth. It also indicates that society and economics are integral parts of the game. Over the past decade, especially with the 'knowledge econ-omy' defined by the Lisbon Agenda for the European community, societal concerns have been explicitly integrated as legitimate research goals. Non-epistemic values, such as competitiveness, commercial profit, sustainability and social justice, are considered to be fully legitimate goals of research agendas, in particular in nanotechnology and syn-thetic biology (Bensaude-Vincent 2009). No longer is scientific research an autonomous sphere ruled by its own intrinsic norms and values. Heteronomous considerations underlie the funding decisions made by public or private bodies, as well as the decisions made by researchers on the laboratory floor.

Accordingly, social scientists can be asked to intervene on the labo-ratory floor in order to assess the capacity of researchers to integrate social considerations in their daily work. This is called 'mid-stream mod-ulation' because it comes after the launching of projects and before the commercial outcomes (long life to the linear model!). It has been experienced in a number of research centres to discuss researchers' deci-sions while they are being made. According to a report published by two 'embedded humanists' who spent several weeks in a laboratory at the University of Colorado in the USA and at Delft University in the Netherlands it is a positive experience of mutual learning: embedded humanists learn to adjust their expectations to what is practically feasi-ble in a laboratory and natural scientists learn to look at their research through different lenses and 'acknowledge the rationality of alternative views and approaches' (Schuurbiers and Fisher. 2009). Although it may be seen as a modest achievement, it shows that ethical issues that are raised 'bottom-up' are more liable to engage researchers and engineers.

Epistemology as a detour towards ethics

Whatever the interest of such discussions about experimental procedures for increasing the ethical and social awareness of researchers,
they do not challenge the research directions more or less imposed by
national research agencies, which are tacitly accepted by researchers on
the laboratory floor. They still rely on the linear model based on the
alleged possibility of separating basic science from applications. Without entering into a detailed epistemological analysis of technoscience,
one major feature has to be emphasised here for our purpose: laboratory
research is, above all, design (Bensaude-Vincent et al. 2011).

In nanotechnology and biotechnology, atoms, molecules and genes
are no longer considered as the ultimate bricks of matter and life.
They are viewed as machines and devices, as potential and promising
solutions to the problems of energy, environment, pollution, cancer,
etc. This does not mean that nanotechnology and biotechnology are
necessarily and primarily application-oriented sciences. The molecular
machines created in the laboratory are much too primitive to be oper-
ated in the messy environment of the open world. Most of them perform
only one task in highly specific conditions. The 'economy of promises'
that presided over their development tends to occult the cognitive
dimension of technoscientific research. In fact, in a series of interviews
conducted with researchers in nanotechnology, many of them confessed
that they were more interested in understanding nature at the nanolevel
than in constructing a quantum computer, for instance.[3] To them, the
promises of technological outcomes were essentially a rhetorical device
for raising research funds. On the laboratory floor, nanotechnology is,
above all, a cognitive activity mediated by technological systems and
by the design of objects. Knowing through making is the ultimate goal.
However, knowledge no longer comes in the form of a theoretical model
representing natural phenomena. Rather, it comes as a 'proof of concept': the demonstration of capabilities to produce a molecular machine
instantiating a process or a phenomenon. The cognitive gain expected
from the design of objects is emphasised by the extensive and quasi-
ritual use of Richard Feynman's quotation 'What I cannot create, I do
not understand', which synthetic biologists use as a banner to characterise what they do. Designing a molecule, a protein, is supposed to
be the best way to understand how nature works. 'There is plenty of
room at the bottom', another of Feynman's favourite quotations, does
not mean that nanotechnologists are investigating the intimate nature
of atoms and molecules. They are interested in their performances, in

what they can do, rather than in what they are made of. Molecules, gene sequences and proteins are redesigned as devices or machines performing specific tasks: rotor, motor, replicator, oscillator, repressilator... Even atoms can be redesigned in opto-electronics for designing meta-materials with new properties and functions.

Ethics of design

If current research is more about designing or redesigning nature than about representing or understanding nature, ethics should be concerned primarily with design, rather than with applications. What sort of ethical issues can be raised by the focus on design in the laboratory?

The ethics of design relies on the assumption that design is not a neutral activity. Beyond the ethical issues raised by products of design at the nanoscale—such as molecular diagnostic chips or brain-enhancing implants—the process of design in itself is value-sensitive (Timmermans et al. 2011). The moral responsibility of designers is often evaluated according to their intentions. For instance, the responsibility of nanoscientists designing graphene can be evaluated according to their cognitive purposes (understanding the properties of a new allotrope of carbon) or the promise of potential applications in electronics. Accordingly, the ethics of nanotechnology has been focused on the decisions made by scientists and engineers along the process of design. This approach of ethics relies on the tacit assumption that design is the projection of human intentions upon material structures, an assumption underlying Eric Drexler's *Engines of Creation* (Drexler 1986). Drexler forged the image of a molecular manufacturer who would pick and place atoms with the help of nanorobots and arrange them according to his mental pattern. This view of design, remindful of the ancient hyle-morphic model (information of a passive matter), is only adequate for ideal machines. It does not work when it comes to the design of real molecular machines. As Richard Smalley and other nano-chemists argued against Drexler, the delegation of the designer's intentions to a nanorobot is only a fantasy because of the forces between atoms and Brownian motion (Smalley 2001; Whitesides 2001). The molecular manufacturer is more like a game-player, seizing opportunities to develop a design strategy.

However, design is an action that cannot be entirely evaluated by the intentions of the designer. In all design, as Peter-Paul Verbeek argued, intentions are both the driving force and the result of the process of design. 'Intentionality is always a hybrid affair, involving human and non-human intentions. [...] Rather than being "derived" from human agents, this intentionality comes about in association between humans and nonhumans' (Verbeek 2008). This general statement is all the more true for design at the nanolevel. Molecular structures afford dispositions that may inspire new projects and transform the initial intention. Design at the nanoscale is a hybrid process, interweaving human purposes and affordances of a given molecule in a very specific context. The term 'affordance', coined by Rom Harré, refers to the performance or service delivered by a thing to its user. It combines generic dispositions instantiating the laws of nature and specific human purposes.[4]

If subject and object cooperate and can hardly be separated, embedded ethicists working on the laboratory floor meet not only moral subjects in their white coats, but also technoscientific objects, such as instruments, molecular machines, electronic networks, etc. Does this mean that the ethical assessment should include such objects?

Decentring ethics

Kant's traditional question 'What should I do?' may not be the most adequate for addressing the ethical issues raised by nanotechnology. Of course, the actions of moral subjects still have to be considered and the scientists' responsibility remains a hot issue. But addressing this issue precisely on the laboratory floor requires a kind of Copernican revolution. Molecular devices should also retain the ethicists' attention, for three major reasons.

First, as they afford something, they suggest a range of actions and behaviours. Indeed, molecular artefacts are not as prescriptive as ordinary macro-artefacts. They do not carry a 'script' prescribing what to do and when (Akrich 1992). Instead, they display a horizon of limitless possibilities that can interfere with a variety of human purposes. Nano-objects offer a wide range of potentials that trigger imagination. It is precisely because nanotechnology is an enabling technology, because nano-objects broaden the scope of potential actions and transformations of the world, that they can be considered as moral agents.

Second, to a certain extent, nano-objects enjoy a relative autonomy and the kind of freedom traditionally required from moral agents. Although they are usually presented as a means to increase our control over natural phenomena, nano-objects have a life of their own. Far from being obedient creatures they are agencies designed for affording unexpected behaviours, for generating surprises. The uncertainty concerning the behaviour of nanoparticles or nanodevices is not just a side-effect. As Jean-Pierre Dupuy argued, the emergence of unintended and unexpected properties is one of the major purposes of their design (Dupuy 2004). If we assume that nano-objects are shaped by this tension between control and emergence, they have to be reconfigured as partners that co-decide what the next step in a technological project should be. Considering the laboratory as a micro-ecosystem, one can extend the partnership relation that Carolyn Merchant assumes in the relationship between humans and nature to humans and laboratory artefacts (Merchant 2003).

Third, nano-objects do not come into being as meaningless objects that would gain value in the future through their technological applications. These objects shed light on natural mechanisms, open up a breakthrough or close off a research pathway: they mean many different things. In the laboratory, they are extremely vulnerable; they depend on various apparatus, on the entire laboratory environment. Their 'life' is entirely in the hands of technicians and researchers who have to take care of them. At the same time their designers are also vulnerable. Research grants and career opportunities of the research team depend on the behaviour and performances of nano-objects. The interdependence between designers and laboratory artefacts calls for a relational responsibility. Objects are integral parts of the moral landscape. Many values and meanings are embedded in them. As hybrids of human projects and natural phenomena, laboratory artefacts are value-laden and not just value-sensitive (such as the techniques of assisted procreation, for instance, which justify the institution of biomedical ethics). All technoscientific objects, according to Javier Echeverria, are overloaded with valuations that inevitably bring conflicts between different value systems and call for an arbiter (Echeverria 2003). Thus, independently from their potential future applications in the society at large, technoscientific objects configure a moral landscape in the laboratory.

Towards a reflective equilibrium

How to cope with this moral landscape? A potential method for bringing ethics onto the laboratory floor is a normative assessment of the commitments underlying the design of technoscientific objects (Bensaude-Vincent and Nurock 2010). It is based on two major assumptions. As mentioned earlier, technoscientific research is emancipated from the work of purification characteristic of positivistic science. Whatever the explicit social values assigned to research programmes, they rely on a 'metaphysical programme' made of unquestioned and non-testable assumptions about nature and society, which drive research along specific directions (Dupuy 2008, 2009). For instance, some programmes in nanobiotechnology engage a view of life as a do-it-yourself kit, which challenges established moral or religious values (Swierstra et al. 2009). Linked to this metaphysical programme, there is also an underlying 'moral agenda' made of tacit concepts, values, informal rules and intuitive norms. For instance, the design of molecular biomarkers for early diagnosis of cancers challenges the clinicians' view of the normal and the pathologic. It also implies specific views of the public health system, with the tacit norm that individuals are responsible for their conditions, for their health, as well as for their diseases or disabilities. There will be no 'responsible innovation' until the norms and values underlying research programmes in nanotechnology are reflected upon.

> Scientists with blinders are precisely what our societies can no longer afford to train, maintain and protect. Our survival hangs in the balance. We need 'reflexive' scientists: less naïve with respect to the ideological dross enveloping their research programmes; but also more conscious of the fact that the science they do rests ineluctably upon a series of metaphysical decisions. (Dupuy 2008, p. xii)

Accordingly, ethicists on the laboratory floor could have a twofold mission: first, to clarify the implicit moral agenda embedded in the design of technoscientific objects; second, to enhance the reflection of scientists on moral aspects by initiating a collective process of deliberation.

For achieving this dual mission, John Dewey's distinction between 'valuation' and 'evaluation' is a helpful conceptual tool. While evaluations are propositions resulting from a conscious and reflexive judgement involving a cognitive activity, valuation is a more affective and impulsive tendency to move toward or away from objects (Dewey 1939, 1958). On the laboratory floor, ethicists have to first disentangle the implicit valuations through informal interviews with scientists and engineers, then to initiate the process of evaluation of the tacit moral agenda for guiding future research through a collective deliberation involving designers and decision-makers, as well as laypeople. The moral representations generated through the collective evaluation may, in turn, become intuitive valuations that, in turn, will have to

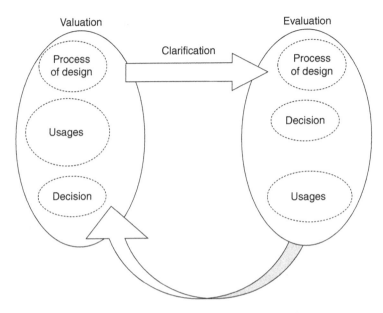

Figure 1.1 A Dewey scheme of deliberation for ethicists on the laboratory floor.

be clarified and evaluated, as suggested by Figure 1.1. By working back and forth among tacit valuations and moral representations, the aim is to arrive at a coherent and acceptable system of values through a process of deliberation, just as in Rawls' 'wide reflective equilibrium' (Rawls 1971).

In this way the moral evaluation is an on-going process accompanying the design of nanodevices centred on the present rather than on speculative futures. Does this mean that the future would not matter, that ethics would be concerned only with our present values or needs without taking care of future generations? Quite the contrary; the reflection on our present valuations opens up new visions of the future. We have seen that the speculative ethics encouraged by ELSI programmes finally took hold of the future. They were future-oriented, but their future was driving them away from the present, from what is actually coming into existence in the laboratories. The future in ELSI programmes was supposed to be the continuation of our present action. It was considered the outcome of our decisions, namely the decision to move forward, to the frontiers of knowledge and technology. This vision of the future is inherited from the modern secular view of time. It is an exemplar of what Barbara Adams named the *future as fortune, i.e. the future in our hands* in contrast to the mythical and religious view of a predetermined destiny, the *future as fate* (Adams and Groves 2007). This modern vision of the future as an empty and abstract space providing a stage for human achievements is still vivid among designers of nanodevices, who seem to forget that the future is never empty. We are the heirs of the so-called empty space of the previous generation's choices for nuclear power, asbestos, etc. A thorough reflection on the futures embedded in the objects being designed as technological solutions to our present problems is important, provided it is informed by our past experiences. The prevalent concept of 'responsible innovation' seems to suffer from amnesia. We have learnt from the past that the future is contingent, that our choices and actions did generate uncontrolled effects. It is precisely because the future lies outside evidence-based science and risk-management techniques that a more reflective attitude is needed. The unbounded exploration of the frontiers of matter driven by an eagerness for incredible and spectacular novelties should give way to a more precautionary attitude. Ethics on the laboratory floor should raise concern about the future of artefacts resulting from creative design. They have a life of their own: How will they evolve out of the confined space of the laboratory? Are they durable, sustainable? What kind of norms and behaviours do they convey? An increased attention to objects of design is a precondition for a collective deliberation on where to invest public money, and which line of research should be pursued or discontinued.

In conclusion, ethics on the laboratory floor is a welcome antidote to the limits of the programmes meant for anticipating the ethical, societal and legal effects of nanotechnology. An epistemological perspective on technosciences, such as nanotechnology or biotechnology, is a useful detour to renew the ethics of emerging technologies. By focusing on the interplay between designers and artefacts, ethics on the laboratory floor looks at objects as moral partners of designers and engineers. On this basis it is possible to configure ethics as an on-going process with feedback between the descriptive assessments of implicit valuations and the normative judgements of evaluation. If ethics on the laboratory floor is a promising approach it is not because it would be more upstream than the ELSI approach and would allow more control over the future; rather, it broadens our view of the future. Instead of a 'not yet' that could be anticipated, the future is the contingent and unknowledgeable 'becoming' of objects that are currently being designed in nanotechnology laboratories and that will operate in the world that we share with them.

Notes

1. See, for instance, the ethics toolkit provided by the European Observatory NANO (http://www.observatorynano.eu/project/catalogue/4ET/).
2. Taking seriously the notion of technoscience and using it to describe today's scientific research does not mean that technoscience would be a new paradigm following an epochal break in the 1980s as Paul Forman, for instance, argued (2007). See also Nordmann et al. (2011).
3. An interview of Christian Joachim, a physicist working on molecular electronics by Bensaude-Vincent and Marina Maestrutti Toulouse, 29 March 2004 : 'Indeed we can claim that we are designing a switch, but it is mainly a subject of fundamental research. Intramolecular electron transfer is a generic subject. It is so magic! Tunnel effect is a bit magic, we still don't quite understand what is going on in the STM [scanning tunelling microscope]... And we want to know more about it'.
4. 'In many cases the outcome of activating a disposition does not depend on any particular human situation, interest, or construction. However, in some cases the phenomenon has a specifically human aspect. Compare the generic outcome that ice of a certain thickness can bear a certain weight per unit area, expressed in a generic disposition, with the claim that ice of that thickness affords walking for a person. [...] The phenomena that are produced are the manifestations of affordances. These are dispositions that bring together two sets of causal powers that cannot be disentangled. There are the powers of the material stuff organized as an apparatus and the powers of the world realized in the phenomena' (Harré 2003, p. 73F).

Further reading

Latour, B. (1999) *Pandora's Hope* (Cambridge, MA: Harvard University Press).

References

Adams, B. and Groves, C. (2007) *Future Matters, Action, Knowledge* (Leiden: Brill).
Akrich, M. (1992) 'The De-Scription of Technical Objects', in Bijker, W. and Law, J. (eds) *Shaping Technology/Building Society: Studies in Sociotechnical Change* (Cambridge, MA: MIT Press).
Bensaude Vincent, B. (2009) *Les vertiges de la technoscience* (Paris: Éditions la Découverte).
Bensaude, V.B. and Nurock, V. (2010) 'Ethique des nanotechnologies', in Hirsch, E. (ed.) *Traité de bioéthique*, vol. 1, pp. 355–69 (Paris: érès).
Bensaude, V.B., Loeve, S., Nordmann, A., and Schwarz, A. (2011) 'Matters of Interest: The Objects of Research in Science and Technoscience', *Journal for General Philosophy of Science*, 42(2): 365–83.
Carrier, M., Howard, D. and Kourany, J. (eds) (2008) *The Challenge of the Social and the Pressure of Practice. Science & Values Revisited* (Pittsburgh: University of Pittsburgh Press).
Dewey, J. (1939) *Theory of Valuation* (Chicago: University of Chicago Press).
Dewey, J. (1958) *Experience and Nature* (New York: Dover Publications).

Drexler, E. K. (1986) *Engines of Creation*. (New York, Anchor Books).

Dupuy, J.-P. (2004) 'Complexity and Uncertainty', in Nordmann, A. (ed.) *Foresighting the New Technology Wave, High-Level Expert Group* (Brussels: European Commission).

Dupuy, J.-P. (2008) 'Foreword: The Double Language of Science, and Why it is so Difficult to Have a Proper Public Debate About the Nanotechnology Program', in Allhoff, F. and Lin, P. (eds) *Nanotechnology and Society. Current Emerging Ethical Issues* (New York: Springer).

Dupuy, J.-P. (2009) 'Technology and Metaphysics', in Kyrre, J., Olsen, B., Pedersen, S. A., and Hendricks, V. F. (eds) *A Companion to the Philosophy of Technology, Blackwell Companions to Philosophy* (Malden, MA: Wiley-Blackwell).

Echeverria, J. (2003) *La revolucion tecnoscientifica* (Madrid: Fondo de Cultura Economica).

European Union High-Level Expert Group (2004) *Foresighting the New Technology Wave, Converging Technologies – Shaping the Future of European Societies* (Brussels: European Union).

Forman, P. (2007) 'The Primacy of Science in Modernity, of Technology in Postmodernity and of Ideology in the History of Technology', *History and Technology*, 23(1/2): 1–152.

Harré, R. (2003) 'The Materiality of Instruments in a Metaphysics for Experiments', in Radder, H. (ed.) *Philosophy of Scientific Experimentation* (Pittsburgh, PA: University of Pittsburgh Press).

Hottois, G. (1984) *Le signe et la technique. La philosophie à l'épreuve de la technique* (Paris: Aubier).

Latour, B. (1987) *Science in Action: How to Follow Scientists and Engineers Through Society* (Cambridge, MA: Harvard University Press).

Merchant, C. (2003) Reinventing Eden; *The Fate of Nature in Western Culture* (New York; London: Routledge).

Merton, R. K. (1986) 'The Ethos of Science [1942]', in Sztompka, P. (ed.) *On Social Structure and Science* (Chicago: University of Chicago Press).

Mnyusiwalla, A., Daar, A. S., and Singer P.A. (2003) 'Mind the Gap: Science and Ethics in Nanotechnology', *Nanotechnology*, 14: R9–R13.

Nordmann, A. (2007) 'If and Then: A Critique of Speculative Nanoethics', *Nanoethics*, 1: 31–46.

Nordmann, A. (2010) 'A Forensics of Wishing: Technology Assessment in the Age of Technoscience', *Poiesis & Praxis*, 7(1): 5–15.

Nordmann, A., Radder, H., and Schiemann, G. (2011) *Science Transformed? Debating Claims of an Epochal Break* (Pittsburgh: University of Pittsburgh Press).

Rawls, J. (1971) *A Theory of Justice* (Boston, MA: Harvard University Press).

Rip, A. (2009) 'Mind the Gap Revisited', *Nature Nanotechnology*, May: 273–4.

Roco, M., Bainbridge, W., and Alivisastos, P. (2000) *Nanotechnology [NT] Research Directions*. IWGN Interagency Working Group on Nanoscience Workshop Report (Dordrecht; Boston, MA: Kluwer).

Royal Society and Royal Academy of Engineering (2004) 'Nanoscience, Nanotechnology: Opportunities and Uncertainties', available at http://www.nanotec.org.uk/ (accessed January 2013).

Sauvage, J.-P. (2008) 'Commentaire sur l'ensemble des contributions', in Bensaude, V., Larrère, B. R., and Nurock, V. (eds) *Bionano-éthique, perspectives critiques sur les bnionanotechnologies* (Paris: Vuibert).

Schuurbiers, D. and Fisher, E. (2009) 'Lab-Scale Intervention', *EMBO Reports*, 10(5): 424–7.

Smalley, R. (2001) 'Of Chemistry, Love and Nanobots, How Soon Will We See the Nanometer-Scale Robots Envisaged by K. Eric Drexler and Other Molecular Nanotechnologists? The Simple Answer is Never', *Scientific American*, 285(3): 76–7.

Swierstra, T., van Est, R., and Boenink, M. (2009) 'Taking Care of the Symbolic Order. How Converging Technologies Challenge our Concepts', *Nanoethics*, 3: 269–80.

Timmermans, J., Zhao Y., and van den Hoven, J. (2011) 'Ethics and Nanopharmacy: Value Sensitive Design', *Nanoethics*, 5: 269–83.

Whitesides, G. M. (2001) 'The Once and Future Nanomachine', *Scientific American* 285(3): 78–83.

Verbeek, P.-P. (2008) 'Design Ethics and the Morality of Technological Artifacts', in Vermaas, P., Kroes, P. A., Light, A., and Moore, S. (eds) *Philosophy and Design: From Engineering to Architecture* (Heidelberg: Springer).

Williams, B. (1981) *Moral Luck* (Cambridge: Cambridge University Press).

Ziman, J. (1994) 'Proprietary, Local, Authoritarian, Commissioned Expert', In: *Ziman Prometheus Bound. Science in a Steady State* (Cambridge, Cambridge University Press).

Ziman, J. (1996) 'Postacademic Science: Constructing Knowledge With Networks and Norms', *Science Studies*, 1: 68–80.

2
Responsible Research and Development: Roles of Ethicists on the Laboratory Floor

Armin Grunwald

Introduction

The rapid and further accelerating pace of science and technology has led to concerns that ethical deliberations often come too late. Ethics in this perspective, could, at best, act as a repair service for problems that are already out in the open: 'It is a familiar cliché that ethics does not keep pace with technology' (Moor and Weckert 2003). In contrast, the 'ethics first' model postulates comprehensive ethical reflection on the possible impact *in advance* of the technological development. It is possible for ethics to reflect and discuss the normative implications of items long before their entry into the market because scientific and technical knowledge will make early ideas available about possible application fields, innovations and their societal impacts (risks, as well as chances). Ethical inquiry already made during the very early stages of a development could provide orientation for shaping either the relevant *process* of scientific advance and technological development, or the envisaged *product lines*. However, ethical reflection in this model has to deal with the control dilemma (Collingridge 1980): the relevant knowledge about technology and its consequences is uncertain and preliminary, a fact that hinders, and possibly prevents, actions of efficiently and effectively shaping technology. In the course of the continuing concretisation of the technological possibilities, it is then possible to continuously concretise ethical assessment and advice by taking into account the continuously improving knowledge base of emerging innovation.

Inevitably, realising the 'ethics first' model includes looking also at what is happening in laboratories and on the 'laboratory floor'. In a metaphor used frequently over the last few years this means 'going upstream' along the development and innovation chain. A new wave of early engagement, sometimes called 'upstream engagement', has been observed in the fields of science and engineering ethics, as well as in technology assessment (TA). This occurred mainly in the field of new and emerging technologies, such as nanotechnology, enhancement technologies and synthetic biology (Grunwald 2012a; Rip and Swierstra 2007). Early ethical exploration and assessment of normative issues of research and development (R&D) has become part of on-going research projects and programmes.

This development can be regarded as one of the origins of the current debate on responsible research and innovation (RRI). The ideas of 'responsible development' in the scientific–technological advance and of 'responsible innovation' in the field of new products, services and systems have been discussed for some years now with increasing intensity. The idea of responsible innovation adds explicit ethical reflection to the 'upstream movement' of TA and incorporates both into approaches to shape technology and innovation. Responsible innovation brings together TA with its experiences on assessment procedures, actor involvement, foresighting and evaluation with engineering ethics, in particular, under the framework of responsibility. Ethical reflection and technology assessment are being increasingly taken up as integrative parts of R&D programmes (Siune et al. 2009).

At this stage of reflection the roles of ethics as a philosophical profession and of ethicists enter the game. Questions arise about the specific responsibilities and tasks of ethicists in the processes of 'upstream engagement'. In particular, looking at the laboratory floor leads to questions, for example, on the cooperation of ethicists with other groups of people who are active there, on division of labour, and on how to ensure the independence of the ethical judgements from the interests of the involved engineers and scientists.

In order to be able to respond to these questions I will first clarify the underlying understanding of responsibility, followed by a conceptual model of the normative uncertainties possibly in place on the laboratory floor. By doing this the field will be prepared for dealing with the roles of ethicists on the laboratory floor, including considerations on chances and risks. By taking cognisance of the fact that R&D is taking place in the laboratories and not innovation, the term 'responsible research and development' (RRD) rather than the RRI notion will be used in this chapter.

Constitutive dimensions of responsibility

The concept of responsibility has been used widely in connection with scientific and technological progress in the last 2–3 decades (Durbin and Lenk 1987). It associates ethical questions regarding the justifiability of decisions in and on science and technology with the actions and decisions of concrete persons and groups and resulting accountabilities, and it is faced with the challenges posed by uncertain knowledge of the consequences of those decisions.

The term 'responsibility' seems, at first glance, to be an everyday word not needing any explanation at all. However, this might be a misleading assumption in the field of science and technology. Often, the notion of responsibility is merely used as a rhetorical phrase with moralistic intention. Making this notion more tangible and operable requires a few clarifying words.

Responsibility is the result of *an act of attribution*, either if actors themselves take over responsibility, or if the attribution of responsibility is made by others. The attribution of responsibility is itself an act that takes place according to *rules of attribution* (on this see Jonas 1979, p. 173). Assignments and attributions of responsibility are made in concrete social and political spaces involving and affecting concrete actors in concrete constellations. They may change the governance of a specific field, and often the explicit reflection on and attribution of responsibility *will* influence the governance and decision-making in that field (see Grunwald 2012b).

Thus, the concept of responsibility shows a *governance dimension* without which its place and role in on-going debate cannot be clarified and which is, therefore, constitutive to that notion. The governance dimension mirrors the fact that the attribution of responsibility is an act done by specific actors and affecting others. Attributing responsibilities must, on the one hand, take into account the possibilities of actors to influence actions and decisions in the respective field. On the other hand, attributing responsibilities has an effect on the *governance* of that field. Relevant questions are: How are the capabilities to act and decide distributed among the persons and groups involved in the field considered? Which social groups are affected and could or should help decide about the distribution of responsibility? Do the questions under consideration concern the 'polis' or can they be delegated to groups or societal subsystems? Which implications and consequences would a particular distribution of responsibility have for the governance of the respective field?

Having seen in this way that responsibility belongs to the social, and possibly political, sphere of a given context it seems quite obvious that responsibility also includes an *ethical dimension*. This is addressed by the question of whether actions and decisions should be regarded as responsible according to the rules and principles determining the *normative* basis of action in the respective field. Insofar as *normative uncertainties* arise (Grunwald 2012a, Ch. 3), for example because of moral conflicts or moral indifference, or because of the insufficiency of the available set of moral rules, ethical reflection on these rules is needed. Relevant questions are: Which criteria allow the distinction between responsible and irresponsible actions and decisions? Which traditions, such as Kantian or utilitarian ethics, should be involved and what would follow out of them? Is there consensus or controversy on these criteria among the relevant actors? Can the actions and decisions in question be justified with respect to the rules, values and ethical principles?

The *epistemic dimension* as the third constitutive dimension of responsibility asks for the quality of the knowledge about the subject of responsibility. This is a relevant issue in debates on scientific responsibility because statements about effects and consequences of science and new technology frequently show a high degree of uncertainty (von Schomberg 2005). The comment that nothing else comes from 'mere possibility arguments' (Hansson 2006) is an indication that in debates about responsibility it is essential to critically reflect on the status of the available knowledge from an epistemological point of view. Relevant questions are: What is really known about prospective subjects of responsibility?; What could be known in case of more research, and which uncertainties are pertinent?; How can different uncertainties be qualified and compared with one another?; What is at stake if hypothetical worst-case scenarios were to become reality?

This brief analysis shows that issues of responsibility are inevitably interdisciplinary, touching upon all of these dimensions. Responsibility is not one of abstract ethical judgements, but entails the observance of concrete contexts and governance factors, as well as of the quality of the knowledge available. RRD must be aware of this complex semantic nature of responsibility. In particular, a cooperation of applied ethics addressing the moral dimension, philosophy of science taking care of the epistemic dimension, and social science (science and technology studies) researching the social and political dimensions, as well as governance issues, is needed.

This threefold nature of the concept of responsibility seems to be particularly relevant to the laboratory context. One argument is that the division of labour and issues of cooperation between laboratory researchers and ethicists need enlightenment in order to make the governance dimension of responsibility transparent. The ethical dimension is touched upon as soon as there might be normative uncertainties and questions emerging from the on-going R&D processes. The epistemic dimension is valid in this context because of the 'upstream' nature of responsibility reflections on the laboratory floor. This nature implies that the subject of responsibility debates—the knowledge about emerging applications, innovations, products and services, as well as about their possible consequences and impacts in a future society—will be highly uncertain or even speculative. Epistemological reflection will be needed frequently to avoid purely 'speculative ethics' (Grunwald 2012a; Nordmann 2007; Nordmann and Rip 2009; Roache 2008) and 'mere possibility arguments' (Hansson 2006).

Responsibility configurations on the laboratory floor

Why, to which ends, and in which cases, ethics and responsibility reflections are needed on the laboratory floor is not absolutely clear per se. In order to identify occasions and procedures to strengthen science's moral responsibility by involving ethicists 'on the laboratory floor' it is necessary to use an explicit and clear notion for modelling the laboratory situation.

In this section I will concentrate on the *ethical* dimension of responsibility because the role of *ethicists* is part of the focus of this chapter. Complementary to, but beyond the scope of, this chapter there could (and should) also be reflections and analyses of the laboratory as a social space (governance dimension) or aiming at enlightening the epistemological issues of possible subjects of responsibility on the laboratory floor.

My point of departure is a more general model of the use of ethics in science and technology (following Grunwald 2012a). The distinction between factual morals and ethics, understood as the reflective discipline for considering moral conflicts or ambiguities, has been widely accepted in the modern discussion. Viewing ethics as a reflective discipline for de facto moral issues means, therefore, that ethical reflection for R&D is required *if the decisions related to the research process and their presumed outcomes involve normative uncertainties*. Often, these uncertainties manifest themselves as moral conflicts about the moral permissibility of the direction, intention and objectives of particular developments, about the means used to achieve progress, and about the moral status of the envisaged outcomes of the development under consideration which might consist of innovations, products, systems and services. Indeed, R&D exhibit ethical aspects in three dimensions: (i) the *goals and purposes* they pursue; (ii) the *instrument* they employ; and (iii) the *consequences and side effects* they will probably produce.

Objectives of research and determining science's agenda

In many cases, the aims of science and technology are morally not problematic. To develop therapies for illnesses such as Alzheimer's disease, to provide new facilities to support disabled people or to protect society against natural hazards visions of this type can be sure to gain high social acceptance and ethical support. In other areas, however, there are ethical conflicts and controversies. The visions related to human space-flight, for example, are controversial in nature, as are research goals for new and more powerful weapons. It is controversial whether human performance should be enhanced and whether research should be done to achieve this. These questions lead to the challenge of 'responsibly' determining the agenda of R&D.

The research process

Instruments, measures and practices of R&D may give rise to ethical questions. Examples are the moral legitimacy of experiments with animals or of practices making use of human persons, embryos or stem cells as subjects of research. Other examples are experiments with genetically modified organisms or plants, especially their release outside of laboratories and, in earlier times, the experimental testing of nuclear weapons. Conflicts regarding the moral status of the human embryo and of animals, as well as questions regarding the acceptability of the risks, regularly lead to severe ethical controversies.

Intended and non-intended consequences

Many of the consequences of science and research are welcome and, without doubt, ethically positive: contributions to health, welfare, employment, development and so forth. These positive consequences must not be neglected in ethical reasoning and assessment. However, since the 1960s, the unintended and adverse effects of scientific and technical innovations have been considerable, and some of them were of dramatic proportions. This situation leads to ethical challenges: How can a society that places its hopes and trust in innovation and progress protect itself from undesirable, possibly disastrous side effects? How can it preventatively stockpile knowledge to cope with possible future adverse effects? Under which circumstances is it responsible to expose people or the environment to risks from technology? How is responsible action possible in view of the high uncertainties possibly involved?

At all stages of R&D decisions have to be made. In some of these decision-making processes ethical and responsibility reflections will be required owing to the virulence of normative uncertainties and moral controversies; in others, it will be possible to make decisions without ethical scrutiny. But how can we distinguish between both types of situations and thus identify the need for ethical reflection?

As a response to this challenge I introduced the distinction between 'standard situations' of moral respect and 'non-standard' situations, distinguished by different relations between the normative framework governing a specific decision-making situation and the decision-making needs (first ideas in Grunwald 2000, and developed through various stages up to the presentation given in Grunwald 2012a). The normative framework includes all regulative and evaluative parts governing acting and decision-making in a specific situation, 'hard' regulations, such as law, as well as 'weak' ones, such as customs, codes of conduct and rules of behaviour. While in a 'standard' situation the normative framework in place provides good, consistent and comprehensive orientation for decision-making, there are conflicts, ambiguities or indifferences of moral respect in the non-standard situation.

Probably the great majority of decisions on R&D in the laboratory may be classified as 'standard' in the following way: the elements of the regulative framework in place are accepted *as given* in the respective situation and they will allow for making a rational decision. In this class of situations no *explicit* ethical reflection is necessary when an engineer in a laboratory thinks over the question of whether she or he should use iron or aluminium for a certain component, when a manager has to decide whether she/he should order production unit A or B for a particular area or when a licensing authority has to decide on an application for building a chemical manufacturing plant—as long as normative uncertainty does not occur. In order to better understand the boundaries and limits of standard situations, the following criteria have been proposed (expanding upon Grunwald 2000, cf. Grunwald 2012a; see also van Gorp 2005, and van Gorp and Grunwald 2009):

- pragmatic completeness
- local consistency
- sufficient lack of ambiguity
- acceptance
- compliance.

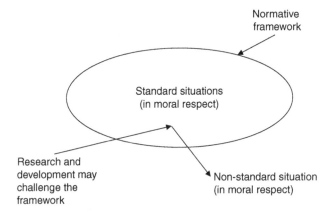

Figure 2.1   The basic model.
(*Source*: Grunwald 2012a, Ch. 3, modified).

If these conditions are satisfied neither moral conflicts nor ambiguities enter the decision-making process in the laboratory. There is, consequently, no need for explicit ethical reflection and thinking about responsibilities. Participants and others affected by a decision can take information about the normative framework in place and the corresponding distribution of responsibilities into consideration as axiological information without having to analyse and reflect on it.

However, R&D may challenge and possibly 'disturb' standard situations of moral respect, transform them into non-standard situations, and make ethical and responsibility reflection necessary (Figure 2.1). The very nature of R&D aiming at providing *new* knowledge and *new* technology is at the core of processes that frequently challenge existing normative frameworks. It is simply the intended novelty of the outcomes of laboratory research that again and again leads to new questions involving normative uncertainties, indifferences, ambiguities and conflicts about responsibility. Often, the established normative framework then no longer provides sufficient and clear orientation on how to proceed. It will then become a matter of debate, inquiry or controversy as to what should be regarded as responsible and what as irresponsible, how the distribution of responsibility should be organised, and which decisions will have to be made involving ethical and responsibility arguments.

In this 'non-standard' situation there are simply three structurally different options to choose from: (i) either stop R&D in the respective field; (ii) modify the direction of research; or (iii) go for a new 'equilibrium', including the development of a modified normative framework.

(i) *The conservative approach*: stop R&D causing moral trouble—renounce its possible benefits and maintain the initial normative framework. As a rule this option is chosen if there are strong, i.e. categorical, ethical arguments against the new technology. An example is reproductive cloning. Such cloning or research on cloning is prohibited in many countries for ethical reasons; in particular, it was banned at the level of the United Nations. Habermas' (2001) argument against interventions in the germ line claims to be a strong argument, but is still an object of controversy.

(ii) *The constructive approach*: attempt to modify some properties of the R&D process that caused moral trouble (maybe circumstances of its production involving animal experiments or the location of a nuclear waste disposal site in a sacred region of indigenous people) in order to be able to harvest the expected benefits without causing moral trouble (see Figure 2.2). The option of *shaping technology* specifically according to ethical values or principles is behind the approaches of constructive technology assessment (see Rip et al. 1995), of the social shaping of technology (Yoshinaka et al. 2003) and of value sensitive design (van de Poel 2009, pp. 1001ff.). The focus is on directing the research, development and design of technical products or systems along the established normative framework so that the products or systems fit into this framework. This would, so to speak, in itself prevent normative uncertainties from arising.

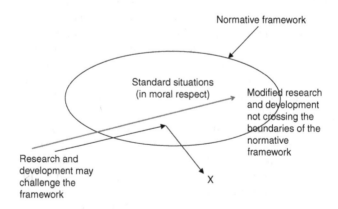

Figure 2.2   Modify innovation.
(*Source*: Grunwald 2012a, Ch. 3, modified).

(iii) *The techno-optimistic approach*: modify the normative framework so that the envisaged technology could be accepted (and the benefits harvested) in a way that would no longer lead to normative uncertainty and moral conflict (see Figure 2.3). This option will enter the debate in cases when the (highly promising) development of a new technology or even research on it would not be possible except by producing normative uncertainty. Examples are animal experiments undertaken for non-medical purposes (Ferrari et al. 2010) or research in which the moral status of embryos plays a role. The issue is then to examine if, and to what extent, the affected normative framework can be modified without coming into conflict with the essential ethical principles. Even the handling of technical risks that have to be tolerated in order to utilise an innovation often takes place by means of modifying the normative framework, such as in the implementation of precautionary measures (von Schomberg 2005).

Responsibility reflections play a decisive role in determining the criteria of choice between such alternatives and—in options (ii) and (iii)—between different versions and according to the concrete context conditions. In these cases the reflection is an act of balancing the expected advantages of R&D and its envisaged outcomes against the moral or

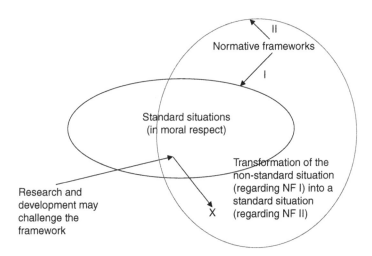

Figure 2.3 Modification of normative framework (NF).
(*Source*: Grunwald 2012a, Ch. 3, modified).

other costs if—as is probably the most common situation—there are no *categorical* ethical arguments for or against.

Roles of ethicists

Close cooperation should help in considerably improving science's responsibility and its ability to establish an 'early warning' function with respect to uncovering possible ethically-relevant questions (in the sense mentioned earlier) emerging from R&D activities on the laboratory floor. Responsibility reflection plays a different role in each of these options, which has implications for the roles of the ethicists involved.

Identification, clarification and treatment

Going through the stages of identification, clarification and treatment of ethical aspects of innovative research, an analytical framework should be developed which could help to focus the 'glasses' through which the division of responsibilities and the interplay of different roles of ethicists and scientists 'on the laboratory floor' could be made more transparent. A number of rough ideas will be presented in this section.

A. Identification

Probably the most challenging task for the entire group in a particular laboratory context will be to discover and uncover situations in which responsibility reflections are required. Assuming that most of the decisions to be taken at that level may be done within a 'standard situation of moral respect' it seems to be a hard task to identify the exceptions from this rule. And it seems to be quite obvious that ethicists have a crucial role in doing this.

Taking over this role of being a 'watch dog' or a 'tracker dog'—it depends on the position as to which notion will be preferred—requires several activities to be undertaken by ethicists on the laboratory floor following a scheme of subsequent tasks.

- The regulative frameworks valid in a specific laboratory context have to be identified as a prerequisite for enabling the group to distinguish between standard and non-standard situations. Because such context conditions, including, in particular, the relevant elements of the normative frameworks governing this context, will not vary rapidly over time it could be the first task of an ethicist entering the laboratory: look for the normative frameworks in place and make them as transparent as possible!

- The range of validity of those normative frameworks is decisive for the standard versus non-standard distinction. Therefore, special attention should be spent to uncovering the possible limitations of their validity. This will help in detecting possible border-crossings in case of pending decisions.
- Having done in this way more or less the necessary preparatory work a careful monitoring of the on-going developments and decisions in the laboratory should be established. In the case of decisions to be made ethicists should prepare judgements as to whether the resulting situation will be located inside or outside of the boundaries of the 'standard situation'. This is the main step for the identification of ethical and responsibility issues.

By completing these steps, and being involved at a very early stage of R&D the entire innovation chain can be considered and exploited. In particular, probably avoiding a situation in which effort, creativity and time would be wasted by researching in a direction leading to severe ethical and responsibility problems, which would possibly prevent the outcomes of that research from having any effect at all.

A major advantage will be the intimate context of ethicists cooperating with natural scientists, engineers, developers and technology designers and, thereby, sharing assessments and judgements, and transferring knowledge across disciplinary borders. The moral awareness and the 'radar' of ethicists concerning 'non-standard situations' in combination with the ideas, knowledge and visions of engineers and natural scientists provide the best possible constellation for 'early warning' with respect to ethical challenges emerging from laboratory work.

However, all persons have to take into account the fact that there will be a situation of high uncertainty with consequences in two directions. First, there is no guarantee that all deviations from standard situations will really be discovered. Some could be overseen because of misleading assessments, inadequate criteria or simply because of unexpected developments. Second, some ethical concern detected by the ethicist might be proven wrong at later stages of development. Both directions of misleading assessment require careful epistemological work on the quality of knowledge which forms the basis of the ethical assessments.

In some situations there might be a moral obligation to go beyond the laboratory context and inform public debate and policy-makers. This will be the case if really new ethical aspects are uncovered that are also relevant to development processes outside the respective laboratory. However, this might be a problematic step because of confidentiality agreements in the contracts—a well-known problem of engineering ethics in general.

B. Clarification

While the identification is a necessary step it is not a sufficient one. In preparing decision-making ethicists can contribute to clarifying the situation. This means that more information should be provided about the normative background of the ethical challenge, including questions such as the following.

- At exactly which point(s) is the respective normative framework in place not sufficient to orientate the situation under consideration? For example, the five criteria mentioned in the section *Intended and non-intended consequences* could be taken as points of departure to clarify, in detail, which of these caused, and for what reasons, the standard situation to be transformed into a non-standard situation. Knowing this could be decisive for determining the type of response to the new situation.
- What fields of ethical inquiry, such as information ethics, privacy issues, safety and security issues, and the rights of living beings, have been touched by the challenge, but are, as yet, uncovered? This question may lead to a focus on the search for solutions in more specific fields that will possibly benefit from work that has already been done.
- Are there novelties in normative or value respect, or are the reasons for the non-standard nature of the situation structurally well-known? In the latter case possibly well-known strategies of problem-solving could be transferred to the situation under consideration while, in the first case, ethical reflection would possibly have to involve research.
- Who is concerned or should be concerned, whose rights might be violated, and which persons and groups would be affected? Are persons or other living beings *today* affected or would it be more of a problem with respect to future generations? Answers to this question could orientate the search for adequate problem-solving strategies and help in identifying existing knowledge.

This role of ethicists is more analytical work and is required as preparation for treatment and problem-solving. In order to be able to fulfil these tasks it will, on the one hand, be fruitful, if not necessary, to be in close contact with researchers from the laboratory. In particular, they will have to provide knowledge about relevant issues of the respective R&D

process, including possible side-effects. On the other hand, knowledge about systems of ethics, ethical argumentation patterns, and fields of applied ethics help to make clearer the issue that is at stake, and knowledge about strategies of ethical reflection and responsibility analyses can be used for problem-solving purposes.

C. Treatment of ethical issues

Finally, decisions have to be made, for example with respect to choosing from the three options distinguished above. While the ethicist will be the expert in performing the two preceding tasks, he or she will only be one member of the group among others in the decision-making process. Ethicists at the stage of decision-making can only provide professional advice, but not replace or dominate the decision-making process (Grunwald 2003).

This observation raises the question as to whose task it will be to make the decision. Generally, answers to this question are related directly to assessments on the distribution of responsibilities and belong, thus, to the governance dimension of responsibility (see above). In modern society, moral labour and responsibilities are distributed in complex arrangements involving the public, the mass media, engineers and natural scientists, philosophers and social scientists, but also stakeholders, regulators and political decision-makers and democratic institutions in different manners according to the respective contexts. What would be needed in this field is a typology of different classes of situations, which could serve as a frame for diagnosing the situation under consideration. For example, laboratories of synthetic biology, which are closely related to basic research and yet far from concrete product innovation for the marketplace (Grunwald 2012b), will need a completely different responsibility scheme compared to product development in informatics of process engineering. Elaborating on this issue, however, is beyond the scope of this chapter.

Coming back to the role of ethicists in the treatment phase shows that, in spite of the more modest direct contribution of ethicists to decision-making according to their place in the laboratory governance, there are specific roles that ethicists could and should take. First of all, consider the three options for responding: the conservative, the constructive and the techno-optimistic option (see earlier). Ethicists are experts for the normative assumptions related to these options. They should act as interpreters as to what these types of options mean and could deliver, so to speak, *hermeneutical* information about these types of options. Second, they could contribute to the inter- and transdisciplinary exchange of perspectives and act as *facilitators* of arriving at a broader picture of the situation under consideration. While engineers and natural scientists frequently stick to the narrower context of their research, ethicists could open up the perspective and demonstrate, for example, that particular aspects of the laboratory research might be related to grand questions such as democracy, division of labour in society, values or images of human nature.

Thirdly, and finally, ethicists could provide a targeted search for alternatives in order to avoid the ethical problems, in particular within the 'constructive' option modifying some aspects of R&D. Obviously, this point also puts emphasis on the necessity of close cooperation between ethicists and the laboratory group.

*   *   *

Processes of R&D cover a broad range of different constellations that lead to different contextual requirements for an optimal involvement of ethics and ethicists on the laboratory floor. Further work should go for a more detailed analysis addressing, if possible, concrete case studies. What can be said at this, more general, level is the hypothesis that among the three activities mentioned earlier the *identification* of ethical issues in early stages will usually have the highest importance of ethics on the laboratory floor. The main role of the ethicist on the laboratory floor is to act as a 'tracker dog' with respect to possible emerging ethical issues at the earliest stage possible. The main arguments for this position are, first, that the identification is the most crucial step of all— without identification and detection of ethical issues there will be no clarification or treatment. Second, and this argument seems to be not that trivial, for the identification the close cooperation of ethicists and researchers is a *conditio sine qua non*. It is exactly this point that seems to be the strongest argument for involving ethicists on the laboratory floor.

Challenges

The three—obviously idealised—roles of ethicists on the laboratory floor contributing to the identification, clarification and treatment of ethical challenges and responsibility problems have common features in that they show an intrinsic need for close cooperation between the ethicists and the engineers, natural scientists, developers and designers. It became clear that there are many potentials of detecting possible ethical challenges as early as possible within this model of involving ethicists directly in laboratory research.

Though many benefits can be expected from involving ethicists on the laboratory floor there are also some concerns. The main problem might be a possible loss of independence as soon as ethical parallel research becomes an integral part of R&D projects. There is a risk of a too-strong identification of the ethicist with the subject of treatment instead of independence and critical distance. Inasmuch as ethics became part of the development process and would identify itself with the technical success, the suspicion might emerge that possible positive results were 'purchased' or that it was nothing but grease in the innovation process. The credibility of ethics—which is essential in order to do its job—would be endangered.

A second critical issue might be freedom of research. In innovation-oriented R&D, strong economic interests are usually part of the game or even dominating, which could lead to conflict. In cases where ethics come up with unwelcome results according to economic interest, voices might be raised to suppress these results. Besides which, free publication will be restricted by confidentiality agreements that would prevent the publication of non-wanted results of ethical analysis. In this case the task of the ethicists would be to convince the partners in the laboratory that problematic ethical results should not be suppressed, but taken seriously in order to re-design R&D in order to avoid problems later on in the marketplace. Problematic results can be mostly reinterpreted as recommendations for modification. In any case, simultaneously sustaining the independence of ethics and its relevance to R&D in concrete R&D processes requires balancing the distance of an observer and the neighbourhood of an involved person, which is an ambitious and delicate task.

These challenges and possible problems should not motivate ethicists to refrain from involving themselves on the laboratory floor. However, strategies of dealing with these challenges have to be established. They could include adequate conflict management, special procedures for quality assurance involving possible external cooperation for the 'purification' of ethical investigation and ensuring the independence of ethical reflection.

Epilogue: Towards responsible R&D

Taking the three dimensions of responsibility seriously and looking at the various roles of ethicists on the laboratory floor leads to the conclusion that RRD unavoidably requires a more intense inter- and transdisciplinary cooperation between engineering, social sciences and applied ethics. The major novelty in this interdisciplinary cooperation might be the integration of ethics (normative reflection on responsibilities) and social sciences, such as STS and governance research (dealing empirically with social processes around the attribution of responsibility and their consequences for governance). RRD—such as RRI (von Schomberg 2012)—is developing into a new and extremely attractive umbrella term with new accentuations which may be characterised by:

- involving ethical and social issues more directly and as early as possible in the research, development and innovation process
- bridging the gap between R&D practice, engineering ethics, technology assessment, governance research and social sciences
- giving new shape to R&D processes and to technology governance according to responsibility reflections in all of its three dimensions mentioned above
- in particular, making the distribution of responsibility among the involved actors as transparent as possible
- supporting 'constructive paths' of the co-evolution of technology and the normative frameworks of society assuming that the 'constructive option' of responding to ethical challenges will be the most attractive one in many cases.

Consequently, RRD harbours a lot of chances and opportunities, but also some challenges. It will be the task of the future research and practice to exhaust the potentials of RRD involving ethicists on the laboratory floor and to establish adequate measures to deal with the challenges.

Further reading

Bijker, W.E., Hughes, T.P., and Pinch, T.J. (eds) (1987) *The Social Construction of Technological Systems. New Directions in the Sociology and History of Technological Systems.* (Cambridge, MA: MIT Press).
Grunwald, A. (2007) 'Converging Technologies: Visions, Increased Contingencies of the Conditio Humana, and Search for Orientation', *Futures*, 39: 380–392.
Grunwald, A. (2009) 'Technology Assessment: Concepts and Methods', in Meijers, A. (ed.) *Philosophy of Technology and Engineering Sciences*, Vol. 9 (Oxford/Amsterdam: Elsevier).
Guston, D.H. and Sarewitz, D. (2002) 'Real-Time Technology Assessment', *Technology in Culture*, 24: 93–109.
Jonas, H. (1984) *The Imperative of Responsibility* (Chicago: University of Chicago Press).

References

Collingridge, D. (1980) *The Social Control of Technology* (New York: St. Martin's Press).

Durbin, P. and Lenk, H. (eds) (1987) *Technology and Responsibility* (Dordrecht/Boston/Lancaster/Tokyo: Reidel Publishing).

Ferrari, A., Coenen, Ch., and Sauter, A. (2010) *Animal Enhancement* (Bern: Neue technische Möglichkeiten und ethische Fragen).

Grunwald, A. (2000) 'Against Over-Estimating the Role of Ethics in Technology', *Science and Engineering Ethics*, 6: 181–96.

Grunwald, A. (2003) 'Methodical Reconstruction of Ethical Advises', in Bechmann G. and Hronszky, I. (eds) *Expertise and its Interfaces* (Berlin: edition sigma), pp. 103–24.

Grunwald, A. (2012a) *Responsible Nano(bio)technology: Ethics and Philosophy* (Singapore: Pan Stanford Publishing).

Grunwald, A. (2012b) 'Synthetic Biology: moral, epistemic and political dimensions of responsibility', in Paslack, R., Ach, J.S., Luettenberg, B., and Weltring, K. (eds) (2012) *Proceed with Caution? – Concept and Application of the Precautionary Principle in Nanobiotechnology* (Münster: LIT Verlag).

Habermas, J. (2001) *Die Zukunft der menschlichen Natur* (Frankfurt: Suhrkamp).

Habermas, J. (2003) The future of human nature (Cambridge/Oxford/Malden: Polity Press).

Hansson, S.O. (2006) 'Great Uncertainty About Small Things', in Schummer, J. and Baird, D. (eds) *Nanotechnology Challenges – Implications for Philosophy, Ethics and Society* (Singapore: World Scientific Publishing), pp. 315–25.

Jonas, H. (1979). *Das Prinzip Verantwortung* (Frankfurt/: Suhrkamp).

Moor, J. and Weckert, J. (2004) 'Nanoethics: Assessing the Nanoscale from an Ethical Point of View', in Baird, D., Nordmann, A., and Schummer, J. (eds) *Discovering the Nanoscale* (Amsterdam: IOS Press).

Nordmann, A. (2007) 'If and Then: A Critique of Speculative NanoEthics', *Nanoethics*, 1: 31–46.

Nordmann, A. and Rip, A. (2009) 'Mind the Gap Revisited', *Nature Nanotechnology*, 4: 273–4.

Rip, A. and Swierstra, T. (2007) 'Nano-ethics as NEST-ethics: Patterns of Moral Argumentation About New and Emerging Science and Technology', *Nanoethics*, 1: 3–20.

Rip, A., Misa, T., and Schot, J. (eds) (1995). *Managing Technology in Society.* (London: Pinter Publisher).

Roache, R. (2008) 'Ethics, Speculation, and Values', *Nanoethics*, 2: 317–27.

Siune, K., Markus, E., Calloni, M., Felt, U., Gorski, A., Grunwald, A., et al. (2009) 'Challenging Futures of Science in Society'. Report of the MASIS Expert Group (Brussels: European Commission).

van de Poel, I. (2009) 'Values in Engineering Design', in Meijers, A. (ed.) *Philosophy of Technology and Engineering Sciences*, Vol. 9 (Oxford/Amsterdam: Elsevier).

van Gorp, A. (2005) *Ethical Issues in Engineering Design; Safety and Sustainability. Simon Stevin Series in the Philosophy of Technology* (Delft and Eindhoven: Simon Stevin Series in the Philosophy of Technology).

van Gorp, A. and Grunwald, A. (2009) 'Ethical Responsibilities of Engineers in Design Processes: Risks, Regulative Frameworks and Societal Division of Labour', in Roeser, S. (ed.) *The Ethics of Technological Risk* (London: Earthscan).

von Schomberg, R. (2006) 'The Precautionary Principle and Its Normative Challenges', in Fisher, E., Jones, J., and von Schomberg, R. (eds) *Implementing the Precautionary Principle; Perspectives and Prospects* (Cheltenham/ Northampton, MA: Edward Elgar Publishing).

von Schomberg, R. (2012) 'Prospects for Technology Assessment in the 21st Century: The Quest for the "Right" Impacts of Science and Technology. An Outlook Towards a Framework for Responsible Research and Innovation', in Dusseldorp, M. and Breecroft, R. (eds) *Technikfolgen abschätzen lehren; Bildungspotenziale tranzdisciplinärer Methoden* (Wiesbaden: Springer).

Yoshinaka, Y., Clausen, C., and Hansen, A. (2003) 'The Social Shaping of Technology: A New Space for Politics?', in Grunwald, A. (ed.) *Technikgestaltung: zwischen Wunsch oder Wirklichkeit* (Berlin: Springer).

3
The Multiple Practices of Doing 'Ethics in the Laboratory': A Mid-level Perspective

Marianne Boenink

'There is not a philosophical method, although there are indeed methods, like different therapies'.

Ludwig Wittgenstein, *Philosophical Investigations*, p. 13

Introduction

When ethicists go to the laboratory they move 'upstream' in the development of science and technologies. This move is often justified by two arguments. The first one hinges on ethics' *effectiveness*: doing ethics in the laboratory presumably creates more opportunities for co-shaping the eventually resulting technology. The second is concerned with the *relevance* of ethics: close cooperation with scientists and engineers may help to focus ethical reflection on the most urgent and pressing ethical issues, and to avoid empty speculation about what might happen in the future. If ethics in the laboratory lives up to these promises, it seems a valuable and justified move, creating an ethics that fits the complex and evolving character of its object: new and emerging science and technologies. Talking about 'ethics in the laboratory' suggests, however, that there is a uniform and well defined way of doing ethics in the laboratory. This is hardly the case; there is a set of publications sharing the idea that it would be useful if ethics enters the laboratory [or, more generally, becomes involved in real-time research and development (R&D)], suggesting various approaches for doing so.

These publications usually start by arguing why it might be useful to go to the laboratory. Subsequently, they present case studies of actual ethical work done in the laboratory, offering proof of principle that going to the laboratory is viable and effective. This structure has several drawbacks. First, the concrete activities performed in the laboratory are usually not described in much detail. They tend to be squeezed in between the theoretical arguments and the presentation of the results, with the results used to justify the necessity of ethics in the laboratory. What is missing, then, is intermediary work reflecting on how to actually do ethics in the laboratory. What exactly is it that ethicists do in the laboratory? And how to do it? Second, because of the self-justifying character of most texts, each article seems to claim that 'this is what ethics in the laboratory should be like'. The different articles hardly comment or build on each other, making it difficult to determine to what extent the suggested approaches exclude or complement each other.

The starting point of this chapter is that it is time to go beyond the self-justifying character of the current 'ethics in the laboratory' literature and to reflect on *the actual practices* of doing ethics in the laboratory. I will distinguish five activities that ethicists actually perform in the laboratory and discuss examples of each. This 'Wittgensteinian' turn to practice may help to understand how ethics in the laboratory consists of several activities, with distinct effects, characteristics and preconditions. It may also help to understand the normative role these activities can have, even when they appear to be descriptive at first sight. Ultimately, this overview and discussion should help future practitioners to determine what activities might suit their goals and the case at hand. I will start with a brief elucidation of the way I selected and read the literature on ethics in the laboratory. Subsequently, I will reconstruct five ethical activities performed in the laboratory. I proceed with a reflection on the 'family resemblances' among these activities, and conclude by explaining why and how this 'mid-level' approach might contribute to doing 'ethics in the laboratory' more generally.

Identifying ethics in the laboratory

Analysing the activities actually performed by authors doing ethics in the laboratory presupposes, of course, some prior delineation of which literature exemplifies 'ethics in the laboratory'. In this chapter, I will work with a broad characterisation of the phenomenon and the two concepts implied. Neither 'ethics/ethicist' nor 'laboratory' will be used in a literal way; I will, nonetheless, continue to use these terms for ease of reading.

For me, the term 'laboratory' in 'ethics in the laboratory' indicates that ethical work engages with and reflects on on-going scientific work and technology development (R&D and technological innovation) in real-time. Desk work, such as the preparation of scientific research proposals, publications and patent applications, is part of the 'laboratory', as are clinical trials or other types of empirical work (partially) performed outside the physical laboratory walls. Moreover, part of the ethical work may be performed in domains different from where scientific research and technology development usually take place, provided that the aim is to feed the observations back to those directly involved in R&D. So interviews or focus groups with potential users of a technology in themselves cannot be considered as 'ethics in the laboratory'; however, if the results are fed back to scientists and engineers, they can be included in it. Ethics 'in the laboratory' in this chapter, then, is interpreted as ethical engagement with the work done in science and technology development, and with the people doing R&D.

What about 'ethics'? Whereas morality is the set of norms and values current in a certain community, ethics is the reflection on morality. Such reflection is not limited to ethicists with a specific academic training; most human beings engage in it from time to time. Moreover, in this chapter doing ethics is not identical to passing judgement on the moral desirability of an act or a way of doing. It encompasses a broad set of activities, including recognising and interpreting the values at stake in a specific situation, imagining how the meaning of these values might shift because of changes in the situation at hand, imagining alternatives for action, as well as the consequences of these actions and their meaning for stakeholders involved. As this book focuses on the question of how to stimulate ethical reflection on technology development during R&D, I conceive ethics here as the broad set of activities performed to reflect on the societal and moral implications of developments in science and technology, initiated by a person *not* belonging to the standard actors in R&D. This can be an ethicist in the strict sense, but also a humanities scholar or a social scientist. So when I use 'ethicist' in the remainder of this chapter, I refer more to a role than to a specific disciplinary background. Ethical work performed routinely by scientific researchers themselves for the purposes of this chapter will not be labelled 'ethics in the laboratory', nor will work by social scientists or humanities scholars that does not have a normative goal.[1]

Circumscribed in this way, it is possible to identify a limited set of literature on ethics in the laboratory. I looked, in particular, at articles that present both general arguments for ethics in the laboratory and actual experiments with such a form of ethics. I then selected a number of articles that displayed different types of ethical activity and, as such, might have proved illuminating for the variety of activities labelled ethics in the laboratory: van der Burg (2009), Fisher et al. (2006), Fisher (2007), Gorman et al. (2004), Lucivero (2012), Rabinow and Bennett (2009), Schuurbiers (2010, 2011) and Zwart et al. (2006). Based on this set of literature, I identified five activities performed while doing ethics in the laboratory. The first four are *diverging* in the sense that they tend to broaden, multiply or open up current ways of thinking and doing in the laboratory. The last one is *converging* in the sense that it brings a broad set of considerations back together, pointing to specific choices or actions.

Before I go on to discuss the first activity, a few clarifications and warnings are provided. First, the distinction between the activities serves analytical purposes. In practice, the five activities are often closely linked to one another. More often than not it would not make sense to perform just one of them. Distinguishing them nonetheless helps to make visible how different activities may achieve slightly different goals. I will discuss the relations between the activities more extensively in the last section. Moreover, although I will often use just one or two publications to illustrate a specific activity, the approach put forward by the authors of these articles should not be identified completely with the activity at hand. Most approaches use more than one activity, and all activities are used by several authors. Finally, the activities discussed here are probably not exhaustive; ethics in the laboratory is a creative and evolving enterprise. The activities discussed here should be seen as a temporary inventory of the type of work done by relative outsiders to stimulate and deepen ethical reflection on emerging technologies during the R&D stages.

'What': Specifying

The first activity I want to discuss is focused on understanding what technology development in a particular laboratory is all about. What exactly is the object that is being developed and what are the objectives aimed at? An ethicist starting in the laboratory will often try to get a good view of both the technology worked on and the values pursued. Asking questions about what is going on, but also further clarifying what exactly is meant by certain phrases ('What do you mean when you say X?'), the ethicist gradually zooms in from a general understanding to a more detailed, specific understanding of the technology being developed. This 'specification work' and the resulting understanding are not simply preconditions for the ethical work to come. The activity of specifying itself contributes to ethics in the laboratory. The question 'what do you mean when you say...', for example, in itself can play an ethical role. This becomes visible, for example, in Lucivero (2012). She engaged with scientists working on what is called 'the nanopill', an ingestible pill with a sensoring system using nano-wires that should be able to detect colon cancer and send a warning to a person's mobile phone at an early stage. This technology is claimed to become an effective, easy-to-use and patient-friendly device compared with existing tools for detecting colon cancer (in particular, colonoscopy). Although the nanopill is definitely not a nearby reality, the vision of such a pill functions as a powerful image when explaining and justifying the research before funding agencies or a general audience. The image of a pill is used in, among others, a video animation on *YouTube*, in a children's book, in many press releases and in interviews given by the laboratory researchers. When entering the laboratory and talking to the scientists and the oncologist involved about what exactly is meant by 'the nanopill', Lucivero learnt, however, that the identity of the nanopill was actually less stable and uniform than it appeared in the public domain. The nanopill was supposed to be ingestible and able to detect colon cancer, but in many further respects it (still) had a flexible identity. The researchers chose a specific biomarker for detecting tumours, for example, but reasoned that others could be added relatively simply later on. Several alternatives were also considered for communicating the result. One would be to make the pill release a blue dye in case of a positive finding, so that the user her-/himself could simply see that something was wrong. The other was to include a radio signalling functionality so that the pill could send the result to a mobile phone, which could be either the user's or the doctor's phone. Depending on these choices, as well as the development of the pump sampling body material from the colon, the size of the pill could be more or less big.

Lucivero thus notes that 'the' nanopill is not one device, but multiple (fictive) ones, and that it would make more sense to talk about nanopill 1, nanopill 2, etc. Such a specification would, first of all, contribute to ethical reflection because one can analyse the implications and assess the desirability of different design options more explicitly and systematically. In addition, specifying brings a broader set of considerations relevant for deciding between options into view, as when Lucivero deconstructs the claimed value of 'patient friendliness'. The different forms of a nanopill would not only quantitatively differ in their realisation of 'patient friendliness', the meaning of patient friendliness would also be slightly different when a blue dye or a radio signal would be used, for example. Specifying the meaning of both 'the nanopill' and 'patient friendliness', then, enables a more systematic analysis and comparison of the overall desirability of different options in technology development. Specification work can be truly effective only, of course, if the different options have not yet been closed down in the R&D process. In addition, laboratory researchers should be willing to include considerations in their decision-making that they did not think of before.

An additional advantage of 'specification work' is that it is particularly well suited in helping to avoid so-called 'speculative ethics'. As Nordmann (2007) and Nordmann and Rip (2009) point out, ethicists reflecting on emerging technologies are liable to go along with the speculative visions put forward by technology developers, irrespective of whether they endorse or criticise these. Such speculative ethics risks irrelevance, may reinforce implausible expectations and possibly neglects more urgent issues in need of ethical attention. Entering the laboratory and asking questions that help to specify the meaning of both the technology in development and the anticipated values helps to clarify what is actually at stake in on-going R&D work. With regard to a technological device, such questions steer a midway between focusing on a clearly envisioned end product and broad visions of technological progress. It helps to see the multiple variants at play in current work on the technology, and how these might affect different values in different ways. Moreover, it helps to bring out the tensions and differences underlying the justifications of a technology in terms of a specific value.

On the one hand, then, specifying is a way to bring into view, discuss and anticipate different, alternative futures of a technology, to juggle with possibilities. Specifying what is actually designed and developed lays bare the uncertainty of laboratory practices, but also the opportunities for changing the course of these practices because of their anticipated (un-)desirability. On the other hand, specifying values is a way to open up the plural meaning of the values referred to by laboratory researchers. If we surmise that controversy about the meaning of values is likely to arise anyway when the technology is introduced into society, advancing it is a way to gain time for addressing it, as well as to prevent clearly undesirable effects.

'How': Reconstructing

A second type of activity found in publications on ethics in the laboratory can be labelled 'reconstructing the grammar of laboratory practice'.

This activity often builds on the specification work described earlier, but it is more concerned with the way of thinking, reasoning and framing (the *how*) in the laboratory than with the object of laboratory research (the *what*). Moreover, the grammar of laboratory practice is often less 'local' than its object, as researchers have acquired it at least partly outside the current laboratory. The grammar of laboratory practice, that is, is often shared with and supported by external actors and institutions. A good example of 'reconstructing the grammar of laboratory practice' is found in Rabinow and Bennett (2009), who were involved with a research consortium in synthetic biology (SynBERC). In an article reporting the first results of their work, they analyse the research programmes operative in SynBERC and synthetic biology more generally. They studied the project proposals and the first publications from a wide set of research projects both within and outside SynBERC. They start by observing that, even though most researchers ascribe the same goals to the field of synthetic biology, in practice these goals are understood in at least four different ways—which is actually a case of *specifying* the objectives circulating in the field. Rabinow and Bennett go beyond specifying, however, to offer a more elaborate reconstruction of the way of thinking and working related to these objectives, distinguishing four 'strategic orientations' in synthetic biology programmes.

Rabinow and Bennett characterise each of the orientations, first on the basis of their principle object of interest. Whereas some synthetic biology laboratories focus on the construction of biological parts, others are more interested in constructing enzymatic pathways, a third group aims at designing and constructing synthetic genomes and a last one focuses on the design of biological systems in their evolutionary milieu. Rabinow and Bennett then specify the type of problems researchers are working on and the goals they aim for. Interestingly, they also point out the analogies underlying the different orientations, as well as some of the conditions for their success. The 'parts'-oriented laboratories, for example, work with the analogy of computer engineering in which standard elements are combined into functional units in a linear way. The 'pathways' orientation, in contrast, draws inspiration from the analogy with industrial chemical engineering, thinking of cells as little factories. Rabinow and Bennett point out the conditions for success in each strategy and the most important venues for realising their aims. They also analyse the limitations of each orientation. Together, these items reconstruct the different ways of thinking and working in laboratories of different orientation. These reconstructions are not meant as a detailed description of what is actually done in specific laboratories (or laboratory groups), even though examples of laboratory groups working in a specific way are given for each orientation. According to Rabinow and Bennett they are Weberian 'ideal types' that sharpen differences and help to characterise emerging trends in science.

As was the case with 'specifying', 'reconstructing the grammar of laboratory practice' is not simply a descriptive activity. Rabinow and Bennett explicitly position their analysis as a contribution to a more responsible form of ethical and human scientific practice, which they call 'human practices research'. They use this term to avoid the orientation on the moral and social consequences of new science and technology in ethical, legal and social issues research, which, according to Rabinow, wrongly implies that humans and society are implied, and should come into view in innovation processes only downstream (Rabinow 2011, p. 159). How does reconstruction help then to improve ethical and social aspects in real-time R&D? Based on their article, the contribution seems twofold.

First, in reconstructing the grammar of laboratory practice, Rabinow and Bennett point out how the specific research orientations implicate and frame human beings and society. Which roles are ascribed to humans, and which ethical or social issues are brought to the fore? The 'parts orientation', for example, puts intellectual property issues explicitly on the agenda, while the 'synthetic genomes approach' has proposed safety-by-design strategies to reduce the risks associated with such genomes when they escape the laboratory. Such a comparative reconstruction brings out the strengths and weaknesses, the resources of, and the white spots in, each orientation with regard to addressing human and social aspects of innovation.

In addition to reconstructing the possibilities and limitations of each strategic orientation from within, Rabinow and Bennett also take a more critical stance by reconstructing the externalities ('the costs expected to be paid by someone else', p. 100) and the critical limitations ('the – often unacknowledged – range of structural capacities and incapacities', p. 100) of each orientation. The safety-by-design approach suggested in the genomes orientation, for example, externalises safety issues that are not amenable to technical design. Its critical limitation is that it is not able to address ethical issues different from such technical safety issues. Thus, Rabinow and Bennett raise ethical questions that most research orientations themselves are not likely to address or are even aware of.

Ultimately, 'reconstructing the grammar of laboratory practices' serves ethical agenda-setting. It lays bare which ethical and social issues are underexposed or outright neglected in laboratory work (but also what is well taken care of). Feeding the insights gained by reconstruction back to the researchers may serve as a catalyst for reflection on, and changes in, R&D practice (as we will see in the next section). Rabinow and Bennett's work also shows, however, that some of the ethical issues may go beyond the laboratory researchers' responsibility. This may motivate the ethicist to move out of the laboratory and to put these issues on the agenda of, for example, policy-makers, funders or citizens in general.

'Why': Probing

The third activity visible in currently published work on ethics in the laboratory consists of probing scientists about their own norms and values: why do they do what they do? This activity is about making fluid, and thus opening up for debate, values and norms that seemed solid—at the level of both individual researchers and research groups. Clearly, this activity can simply be a follow up of reconstruction, when the results of the reconstructive analysis are fed back to specific researchers. However, it can also be a relatively independent activity, not based on prior analysis of the grammars used in a field. A good example of this non-prepared form of probing can be found in the work carried out by diverse groups of social scientists, philosophers and ethicists under the label 'midstream modulation' (Fisher et al. 2006). Midstream modulation is a framework for laboratory intervention that aims to 'elucidate and enhance the responsive capacity of laboratories to the broader societal dimensions of their work' (Schuurbiers 2010, p. 71). The approach thus targets the reflective capacity of scientists and engineers, but with a specific focus on the social and ethical aspects of laboratory work. By asking questions for clarification (What are you doing? How does it work?) and justification (Why are you doing this, in this specific way? Could you have acted differently?), the modulator forces (albeit in a gentle way) the researchers to voice tacit knowledge and assumptions, and to reflect, in particular, on the meaning and legitimacy of assumptions related to values and norms. To be sure, the midstream modulation approach combines specification, reconstruction and probing, without proposing an ordering of these tasks in time. I will highlight here mainly the probing activity.

Schuurbiers (2010, 2011) carried out three-month-long laboratory engagements at two locations (in the Netherlands and the USA) with the specific aim 'to enhance lab-based critical reflections on the broader socio-ethical context of their research' (Schuurbiers 2011, p. 772). Schuurbiers observed laboratory work and laboratory interactions, and interviewed individual researchers, 'routinely asking different kinds of questions than those usually encountered in the midst of laboratory research' (2011, p. 777). He also used a 'protocol' developed by Fisher (2007) for socio-technical integration research (STIR) to map and jointly reflect on the decisions made by researchers. This protocol distinguishes four aspects of decision-making: opportunities, considerations, alternatives and outcomes. Using this tool in repeated conversations with laboratory researchers according to Schuurbiers helps to make unarticulated goals, values and other considerations explicit. It also creates opportunities to reflect upon them. Schuurbiers reports that the topics most often discussed during his laboratory engagement were related to research goals and to scientific techniques. These apparently 'technical' topics, however, also brought into view norms and values widely accepted in the laboratory or the research community, and opened these up for discussion. In addition to these 'micro-ethical issues', so called macro-ethical considerations came to the fore, related to the social responsibility of scientists and engineers, as well as the relation between science, technology and society.

The interaction with Schuurbiers made more visible to the researchers when and how their work is suffused with assumptions about values and norms. Moreover, the enhanced visibility of such values and norms opened them up for reflection and contestation, that is it made them more fluid and liable to change. In the case of micro-ethical assumptions, for example about the need to respect safety and health regulations in the laboratory, the interventions by Schuurbiers did, indeed, lead to change. This was, perhaps unsurprisingly, not the case for the so-called 'macro-ethical issues'. Change at this level, after all, requires much more than individual action. Nonetheless, reflection and discussion on, for example, the societal benefit of research, made researchers more aware of the background theories and value systems underlying their scientific practice—an awareness that might feed into the set up and conduct of their next projects.

Repeatedly asking *what* and *why* questions about the daily activities going on in the laboratory, then, helps researchers to zoom out of their 'business-as-usual' mode, and to start (or restart) thinking about the underlying values and norms. Why do we do it this way and why is that important? What's the value involved? And how does this value relate to other values at stake in the laboratory work? This type of activity may help to investigate whether current ways of doing laboratory work are, indeed, the best way to reach the aimed-for goals. In addition, it can lead to more radical reflection on the desirability of these goals, as well as the context necessary to enable laboratory researchers to contribute to their realisation.

'With and for whom': Broadening

Whereas the preceding activities of reconstructing and probing may help researchers to reconsider their own ways of thinking and working, the activity I will call 'broadening' explicitly aims to confront laboratory researchers with considerations from sources outside the laboratory. The ethicist collects information on relevant practices outside the laboratory, and gathers considerations and views from actors and stakeholders who might be affected by the results of the laboratory work. The results of this explorative work are then fed back to the laboratory researchers. This can be done directly, by organising one or more interactive workshops where stakeholders are brought together with laboratory researchers, or indirectly, when the ethicist acts as a spokesperson for others' concerns.

An example of such broadening work can be found in Zwart et al. (2006). They describe how they charted the network of human actors involved in, and the stakeholders potentially affected by, a novel wastewater treatment technology. They invited representatives of these groups for a session in a so-called group decision room. The goal of this session was to trace agreements and disagreements between stakeholders' risk assessments of the new technology, and to explore how to deal with these risks. First, participants were asked to list the risks associated with the new technology, and then to collectively select the most important ones. Subsequently, all participants scored these risks in terms of probability and impact. During the last part of the session they discussed when these risks should be dealt with: during the research, design or use stage of the innovation process. Analysing the session, Zwart et al. observed several interesting divergences and gaps in stakeholders' risk assessments. They point out, for example, that users put more stress on the manageability of the technological process, that is on what is necessary 'to keep the process running whatever the fluctuation in influent', whereas researchers tended to focus more narrowly on the controllability of the process, 'in the sense of knowing the relevant mechanisms and input variables that determine the output variables' (p. 669). As a result, the researchers anticipated fewer problems than users. Moreover, Zwart et al. observed that all actors delegated the responsibility for risks of secondary emissions to a phase of the innovation process for which they were not primarily responsible, thus 'orphaning' this specific risk.

The authors then go on to explain these observations in terms of the roles, responsibilities and agendas for the specific stakeholder groups involved, pointing out which ethical issues need further exploration and discussion. As Zwart et al. claim, such a broadening exercise helps 'to discern ethical issues that would otherwise probably have been over-looked, by the actors and philosophers alike' (p. 681). Thus, the effect not only concerns researchers' thinking, but other stakeholders' views and the ethicists' analysis are also broadened.

Broadening, like the preceding activities, contributes to ethical agenda-setting. In this case, however, the agenda is a starting point for *collective* learning by a number of stakeholders. As a result, it is an open question as to whether the conclusions of this learning process will be relevant for the laboratory researchers or whether they need to be taken up by other actors. Moreover, the format of broadening exercises may differ greatly. The session in the group decision room described by Zwart et al. was very much prestructured by the ethicists who were also the ones interpreting the results. It is possible to have sessions in which the ethicist's role is more that of a facilitator (as described, for example, by Krabbenborg in Chapter 9). In contrast, the ethicist could also opt for a stronger, mediating role. Van der Burg (2009) describes, for example, how an ethicist imagines the effects a technology-in-development might have beyond the laboratory by way of (i) making researchers' scenarios of future use explicit (actually a form of specifying their visions); (ii) research into the history of the technology and current practices of use; and (iii) interviews with stakeholders. The resulting considerations are voiced and fed back to the laboratory by the ethicist her-/himself.

It is important to note that broadening, in whatever form, presupposes that it is clear where to look for the relevant input for technology developers' thinking. This applies to the selection of stakeholders, as well as to the focus that is part and parcel of any format chosen. Such selections can be particularly difficult because emerging technologies are, by definition, 'in the making'. For a start, the ideas about the material 'device to be' may be hazy at best, different options may be circulating at the same time, or work may focus on functionalities without reference to a specific device or context of use. An emerging technology can also create completely new practices of use, as well as new user or stakeholder groups. All this need not necessarily block or hinder the relevance of broadening technology developers' thinking, but it implies that ethicists should take care in choosing where to go for broader input. To complicate things further, broadening in principle is an activity without a clear end point, whereas time (and researchers' interest) is usually limited. It is up to the ethicist, then, to judge what the most sensible broadening move in a particular situation is.

'Whither': Converging and aligning

As indicated earlier, the preceding four activities all have a *diverging* or multiplying effect on laboratory researchers' thinking and doing. To result in improved choices and actions, however, the diverging activities need to be accompanied by a *converging* moment, when the multiplicity of viewpoints and considerations are condensed into specific decisions and/or acts. Current literature on ethics on the laboratory floor, however, is surprisingly silent on how convergence actually occurs and which role the ethicist can play in it. Some authors stop short of describing any change in decisions or actions, suggesting that broadening thinking and options for actions is in itself a sufficient result. Others do describe that researchers actually changed their course of action after the ethical activity in the laboratory, but it often remains mysterious what the ethicist's role in this stage actually was. I will discuss two examples that show, at least to some extent, what may be going on in this step.

Fisher (2007) relates how a doctoral student, by engaging in reflections with an ethicist, came to a sequence of decisions based on the broad set of technical, societal and pragmatic considerations brought out during these reflections. Fisher describes the moment of convergence in terms of 'cognitively linking (societal) considerations with (technical) alternatives' and as 'the alignment of non-research-normal components with research-normal components' (p. 163). Although the article does not discuss in detail what is required to bring about such an alignment, it does give some clues. First, Fisher stresses that the STIR-protocol (discussed earlier) enabled the researcher to become aware of considerations as well as alternatives. The ethicist did not put forward specific 'ethically relevant' considerations, but facilitated changes in decision-making by probing the researcher about the 'what and why' of his thinking and doing. Ultimately, this led the researcher to reflect on (among others) the environmental and health effects of the material commonly used in a specific experiment. However, decision-making was affected only after the researcher reframed the social considerations (what would be best for the environment) in technical terms (what are the materials available). Finally, the narrative brings out the inherently creative nature of identifying and aligning very different considerations. The moment of convergence depended partly on the doctoral student's own perceptiveness for what could be relevant and on his capacity to imagine alternative options.

This is different in Gorman et al. (2004), who relate how a materials scientist and a social scientist jointly supervised a graduate student with the aim to 'stimulate cutting edge research directed towards a socially beneficial outcome' (p. 63). The project was set up as a cooperative endeavour, a so-called 'trading zone' between different disciplines. The social scientist's role appears to have been much more substantial than in Fisher's case. He continuously reminded the others of potential societal and ethical considerations, in particular when framing the research goal and questions. Thus, Gorman did not only facilitate the researchers' perceptive and imaginative capacities, he actually added considerations and contributed to their alignment in the final research decisions (to such an extent that he was included in a patent application resulting from the research).

Interestingly, neither Fisher's nor Gorman's role fits with the image of the ethicist offering advice or passing judgement on what is the most responsible way to proceed. There are, indeed, good reasons to be careful with passing explicit ethical advice. First, as the researcher is the one doing laboratory work, it is important that she/he is motivated to act in a responsible way. Such motivation is definitely stronger when she/he arrives at a certain conclusion her-/himself, instead of being advised or prompted by an ethicist. Second, if good ethical judgement is context sensitive (as I would claim it should be), it requires a thorough knowledge of, and experience with, the situation in hand, which the ethicist (even after prolonged stay in the laboratory or interaction with researchers) may not have. Finally, laboratory researchers cannot be held responsible for everything. The role responsibility of laboratory researchers is limited, and ethicists should take care not to burden researchers with misdirected moral expectations.

This being said, the question remains: What is the added value of a more modest ethical role, as displayed in both Fisher (2007) and Gorman et al. (2004)? Of course, a controlled comparison of situations with and without ethical involvement is impossible. The authors mentioned seem to have two reasons for assessing their own role positively. First, the actors involved (the laboratory researchers) indicated that the interaction with the ethicist had shaped and shifted their way of thinking and doing. The second, more implicit, reason is that a broader variety of considerations was taken into account. The question as to whether such deliberation was broadened *enough* and whether the final alignment did *sufficient* justice to all relevant considerations for it to be ethically legitimate is hardly asked. Neither Fisher nor Gorman distinguishes between actual acceptance and ethical acceptability.

Overall, then, the activity of converging and aligning all relevant considerations is rather underexposed in the ethics on the laboratory floor literature. Both the creative moment of converging all considerations and the normative criteria used to assess a decision as an ethical improvement deserve further clarification. I am not saying that we need to elaborate a theoretical framework to guide decision-making and to delineate in advance what is good and bad. But it might be helpful to reconstruct more explicitly how considerations converged into decisions in specific cases, and when and why changed decisions should be considered an ethical improvement. Such insights could then be used to inspire and refine subsequent ethical work in the laboratory.

Family resemblances (and differences)

I started by claiming that the literature on 'ethics in the laboratory' is understood best as a set of articles showing family resemblances. The publications have a lot in common, but they also show many differences. By studying different examples of doing ethics in the laboratory, I teased out five different activities displayed in 'ethics in the laboratory' projects: (i) specifying technological options and values at stake; (ii) reconstructing ways of scientific and technological thinking and framing; (iii) probing laboratory researchers' thinking; (iv) broadening their thinking in confrontation with other stakeholders' considerations and views; and (v) converging the set of legitimate considerations of stakeholders into choices and actions.

These activities might be related to one another as subsequent stages in time, constituting a continuum of steps to go through to do ethics in the laboratory. It seems to makes sense first to become familiar with laboratory practice and researchers' reasoning by specifying and reconstructing, before putting implicit assumptions up for discussion in probing and broadening, while wrapping up by converging and aligning. It would be a mistake, however, to frame the activities in such a linear way *only*. One might, for example, start with broadening right away if this seems a more promising way to make researchers reflect on their assumptions than reconstructing those assumptions yourself. And specification work may be quite urgent at the stage of aligning, even when the ethicist has done such work already when starting her/his laboratory engagement. So, even though it often makes sense to perform the activities as subsequent stages, jumps, feedback loops and re-orderings should be allowed for. What the best combination and order of activities is depends, ultimately, on context.

Such contexts, as the case studies show, vary widely. The trigger for starting an 'ethics in the laboratory' project, the characteristics of the laboratory and the researchers involved, their willingness and possibilities to cooperate, the public and political context, and the limitations of time and funding available, as well as the characteristics and goals of the ethicist her-/himself can be very different from one situation to another. Ultimately, such conditions shape the possibilities for, and the goals of, doing ethics in the laboratory. In view of this variety, a 'one size fits all' proposal for doing ethics in the laboratory would be misguided. Any ethicist considering a project in a laboratory will have to exercise a lot of diagnostic skill to determine which activity (or, rather, combination of activities) best suits the needs of the situation at hand at a specific moment. This also implies that, in some cases, an offer or initiative to join a laboratory should be rejected or cancelled, for example if the situation does not allow for a fruitful performance of any of the activities discussed.

To help future ethicists who plan to do laboratory work to fit their activities to the case at hand, I will end this chapter with a brief overview of the family resemblances and differences among the activities I have been discussing. These are summarised in Table 3.1. I will also provide some brief observations on the strengths and weaknesses of particular characteristics.

1. First, the activities differ as to *the ethical capacity they affect.* Several activities (in particular, specifying and reconstructing, but also probing and broadening) seem to be directed mainly at the clarification and interpretation of the ethical issues at stake (ethical agenda setting). Probing and broadening are also directed at improving deliberation, within the laboratory and in discussion with external stakeholders respectively. Converging, in contrast, is directed at decision-making. Many ethics on the laboratory floor exercises seem to be satisfied with raising ethical issues within the laboratory. They do not explicitly work towards improved decision-making. For a more far-reaching and lasting improvement of laboratory practice, however, deliberation and decision-making are also required. If, as is probably often the case, this is not feasible because of practical limitations, ethicists could at least do more to point out what type of follow up of the issues raised would be desirable.

Table 3.1 Activities that characterise ethics on the laboratory floor

	Effect(s)	Ethical capacity affected	Targeted group	Sources used	Impact on thinking	Role of the ethicist	Relations to researchers
Specifying	Improving understanding of technological options, values at stake, and potential value conflicts and controversies	Agenda-setting	Researchers	Internal (interviews, also documents)	Diverging	Facilitating	Communicative
Reconstructing	Improving understanding of scientific way of framing and reasoning; laying bare strengths and weaknesses of how humans and society are implicated	Agenda-setting	Researchers, but also policy-makers and society at large	Internal (research proposals, publications; interviews)	Diverging	Active, substantial	Communicative
Probing	Triggering reflection on routine ways of doing and thinking in the laboratory	Agenda-setting, internal deliberation and decision-making	Researchers: individuals and as a group	Internal (observations, interviews, often informal)	Diverging	Facilitating	Communicative

Table 3.1 (Continued)

	Effect(s)	Ethical capacity affected	Targeted group	Sources used	Impact on thinking	Role of the ethicist	Relations to researchers
Broadening	Making researchers aware of other stakeholders' considerations/views and have them reflect on their own; collective learning of actors in- and outside the laboratory	Agenda-setting and external deliberation	All stake-holders	External (observations of relevant practices; interviewing or workshops with stake-holders)	Diverging	Facilitating and/or substantial	Communicative and/or collaborative
Converging	Arriving at ethically-improved choices and (plans for) actions	Decision-making	Mainly researchers, with reference to all stake-holders	Both internal and external	Converging	Facilitating and/or substantial	Communicative and/or collaborative

2. The activities also vary with regard to *the group they are addressing*. Whereas all ethics in the laboratory ultimately targets laboratory researchers, some activities (in particular, reconstructing) branch out to policy-makers, funding organisations or citizens, in particular when the issues raised cannot be tackled within the laboratory. Limiting oneself to one clearly identifiable group may be convenient for practical reasons and thus easier to realise. But it risks neglecting that the thinking and doing of laboratory researchers is partially determined by others. So it seems important to at least keep open the possibility that the ethical issues raised should be put on the agenda of groups outside the laboratory.

3. The *sources used as input* for the activities can be internal or external to laboratory practice. In the case of specifying, reconstructing and probing they are mainly internal, such as scientific documents (research proposals, publications) and interviews (whether more or less formal). In the case of broadening, and possibly also when converging, the ethicist explicitly uses sources external to laboratory practice (such as interviews with stakeholders or observations of current practices) to trigger ethical reflection and deliberation. Choices in this dimension will usually be partially driven by practical considerations: what is available or possible?

4. As indicated earlier, most activities have a diverging *effect on laboratory researchers' thinking*, but the last activity has a converging effect. Diverging activities can be very useful to make sure the agenda for ethical debate is sufficiently broad and that the ensuing deliberation is rich enough. However, without subsequent convergence, the diverging activities may just cause temporary stirring of laboratory practice.

5. In a similar vein, the *role of the ethicist* can be substantial or more procedural. When specifying and probing, she/he seems mainly questioning and facilitating. This also means that the lead is given to the laboratory researchers and their responses, which may limit the possibilities for critical distance. This is different while reconstructing and converging, where the ethicist can have a substantial contribution, steering the ethical agenda setting and the actual decision-making process. In broadening, both a substantial and a more facilitating role are possible.

6. The activities also differ as to *the relation to the scientist* necessary to perform them. The relation in specifying, reconstructing and probing seems mainly communicative, exchanging information and reasons. In broadening and, in particular, in converging, the relationship can

be much more collaborative in the sense that both the researcher and the ethicist have a substantial contribution. Choices in this dimension will, again, at least be partially driven by practical opportunities and limitations.

Conclusion

By focusing on the current practices and activities of doing ethics in the laboratory, I tried to harvest experiences of doing ethics in the laboratory, while, at the same time, moving away from the idea that there is just one right way of doing ethics on the laboratory floor. Existing practices of ethics in the laboratory are, indeed, varied; there is no essence or core characteristic beyond my initial characterisation of it as 'activities to stimulate reflection on the societal and moral implications of on-going R&D, by a person not doing R&D herself'—a delineation that is hardly helpful when considering how to go about an 'ethics in the laboratory' project. I re-described current practices in terms of the activities performed, thus at least enabling future ethicists to distinguish different parts of the work and to grasp how they might be ordered and combined. Charting the characteristics of these activities in terms of family resemblances, moreover, helps to determine which activities may or may not be useful and/or feasible in a specific situation.

A central lesson to be derived from this overview is that, as indicated earlier, an ethicist who aims to engage with R&D work should exercise a lot of *diagnostic skill*. The capacity to find out what is necessary, when, where, with whom and how is crucial. What are the most pressing issues at stake, which activities could help address those and who should be involved? Moreover, what is possible in the laboratory or the R&D group at hand, and in view of the time, financial and possibly other constraints posed on the ethical work? Ideally, an ethicist should be flexible in accommodating the differences between cases. This means that she/he should be familiar with, and skilled in, all activities instead of working from a pre-established approach.

To further elaborate and improve on such a 'situationist' ethics in the laboratory, it does not suffice to learn by doing and simply go from one project to another. Continuous attention to, and reflection on, experiences is required, on an individual, but also on a collective level. The family of ethical activities on the laboratory floor would definitely profit from more 'mid-level' work: taking stock of what has been learned already and making sure that others take advantage of earlier experiences, but also reflecting on what is less well developed or outright missing. The lack of good examples of converging and aligning noted above is a case in point. Framed on a general level, two types of mid-level work seem desirable. First, results achieved by doing ethics in the laboratory, as well as insights gained on what hinders laboratory practice from being fully socially and morally responsible, should be brought together. This can help to diagnose when and where a specific form of doing ethics might be necessary and likely to succeed (or not). It might also help identify needs for institutional or political changes in the research system. Second, methodological lessons need to be pooled. Which type of activity was applied when, where, how and with what result? In both cases, not just success stories, but also those of failure should be shared. What is needed, then, is a (virtual) space for regular and honest exchange, reflection and mutual learning—a laboratory of our own.

Notes

1. I excluded, for example, literature on constructive technology assessment (CTA), because CTA does not necessarily have a normative goal. However, many of the activities described here are quite close to the activities performed in CTA exercises. Although it would be interesting to investigate the commonalities and differences between CTA and the set of literature discussed here, this is beyond the scope of the current chapter.
2. In their 2009 paper, Rabinow and Bennett do not discuss whether (and, if so, how) their reconstructive activities were fed back to the researchers working in SynBERC. By publishing it in the journal *Systems & Synthetic Biology*, however, they claim implicitly that their reconstruction might be helpful to researchers in this field, as well as to ethicists wondering how to engage with synthetic biology. In later publications (e.g. Rabinow 2011; Rabinow and Bennett 2012) it becomes clear that the authors became quite frustrated with the lack of response by SynBERC researchers to their work. By the end of 2011, the cooperation was terminated.

References

Fisher, E. (2007) 'Ethnographic invention: probing the capacity of laboratory decisions', *Nanoethics*, 1: 155–65.

Fisher, E., Mitcham, C., and Mahajan, R. (2006) 'Midstream Modulation of Technology: Governance from Within', *Bulletin of Science, Technology and Society*, 26: 485–96.

Gorman, M.E., Groves, J. F., and Shrager, J. (2004) 'Societal Dimensions of Nanotechnology as a Trading Zone: Results from a Pilot Project', in Baird, D., Nordmann, A., and Schummer, J. (eds) *Discovering the Nanoscale* (Amsterdam: IOS Press).

Hymers, M. (2010) *Wittgenstein and the Practice of Philosophy* (Peterborough: Broadview).

Lucivero, F. (2012) 'Too Good to be True? Appraising Expectations for Ethical Technology Assessment', PhD thesis (University of Twente).

Nordmann, A. (2007) 'If and Then: A Critique of Speculative NanoEthics', *Nanoethics*, 1: 31–46.

Nordmann, A. and Rip, A. (2009) 'Mind the gap revisited', *Nature Nanotechnology*, 4: 273–4.

Rabinow, P. (2011) *The Accompaniment. Assembling the Contemporary* (Chicago: University of Chicago Press).

Rabinow, P. and Bennett, G. (2012) *Designing Human Practices. An Experiment With Synthetic Biology* (Cambridge: Cambridge University Press).

Rabinow, P. and Bennett, G. (2009) 'Synthetic Biology: Ethical Ramifications', *Systems and Synthetic Biology*, 3(1): 99–108.

Schuurbiers, D. (2010) 'Social Responsibility in Research Practice. Engaging Applied Scientists with the Socio-ethical Context of their Work', PhD thesis (University of Delft).

Schuurbiers, D. (2011) 'What Happens in the Lab: Applying Midstream Modulation to Enhance Critical Reflection in the Laboratory', *Science and Engineering Ethics*, 17(4): 769–88.

van der Burg, S. (2009) 'Imagining the Future of Photoacoustic Mammography', *Science and Engineering Ethics*, 15(1): 97–110.

Zwart, S. D., van der Poel, I., van Mil, H., and Brumsen, M. (2006) 'A Network Approach for Distinguishing Ethical Issues in Research and Development', *Science and Engineering Ethics*, 12: 663–84.

4
Technology Design as Experimental Ethics

Peter-Paul Verbeek

Introduction

Doing ethics of technology has become a complicated activity following the developments that have taken place in the philosophy of technology over the last few decades. Contemporary approaches, such as actor–network theory and postphenomenology, have argued convincingly that we need to blur the boundaries between human beings and technological artefacts. While Don Ihde has shown that human relations with the world are fundamentally mediated by technologies (Ihde 1990), Bruno Latour claims that we need to give up the separation we make between human and non-human beings (Latour 1993). In the meantime, according to various authors, even the field of morality has become a hybrid affair. Moral actions and decisions of human beings are fundamentally mediated by technologies, like turnstiles that intervene in fare-dodging in the subway and antenatal diagnostic technologies that inform moral decisions about abortion (Verbeek 2011).

This blurring of the boundaries between humans and technologies is a serious challenge for the ethics of technology. It makes it impossible for ethicists to play the role that is typically associated with them: the role of a border guard assessing whether a technology is morally acceptable or not. Ethics can no longer defend a boundary between humans and technologies when the recent insights from philosophy of technology show that this boundary does not exist.

This does not imply, however, that the ethics of technology has reached its end. As I explained elsewhere (Verbeek 2010) the real challenge is to develop new ways of doing ethics of technology, that shift their focus from 'assessing' technologies towards 'accompanying' their development, implementation and use. Rather than determining whether a technology is morally acceptable or not, the ethics of technology could focus on the question of helping to shape good hybrids. And rather than taking on the role of an external judge it then plays the role of an engaged participant who brings in perspectives that might otherwise remain under-represented.

In this chapter I will elaborate what such an 'ethical accompaniment of technology' could entail. First, I will further articulate the notion of 'ethical accompaniment' as opposed to 'technology assessment'. By discussing Michel Foucault's concept of 'limit attitude', I will show that the notion of 'limit' can be used not only to assess the desirability of technologies but also to accompany their development. Following this, I will propose a framework to read and to design the moral significance of technologies. By linking the approach of technological mediation to design thinking, it becomes possible to expand the realm of ethics from words and ideas to things and technological systems. In three sections, I will indicate three ways in which designers could take the mediating role of technology into account in their work: they can *anticipate* mediations, systematically *assess* them and deliberately *design* them 'into' a technology.

Ethics: From assessment to accompaniment[1]

Implicit in many ethical approaches to technology is the model of a struggle between human beings and technologies. While some technological developments can be beneficial, others pose a threat to humanity. The role of ethicists, then, is to assess if technologies are morally acceptable or not. As indicated in the introduction to this chapter, this model of a struggle between humanity and technology has become highly problematic in view of the recent developments in philosophy of technology. According to recent insights, the human being cannot be understood in isolation from technology, just like technology cannot be understood in isolation from humanity. Approaching their relation in terms of struggle and threat is like attempting to give a moral evaluation of gravity, or language. It does not make much sense to be 'against' gravity or language, as they simply form the basis of our existence. In similar ways, technology inevitably helps to shape what it means to be human.

To be sure, this does not imply that all roles of technology in human existence are equally desirable, and that human beings are, in fact, powerless victims of the power of technology. But it does imply that the 'opposition model' of humanity and technology might not be the most productive model if one wants to change undesirable configurations of humans and technologies. Ethics should not focus on defending the boundaries between humanity and technology, but on governing their intertwinement.

If there is a struggle between humans and technologies at all, it should be conceptualised in the way Heidegger conceptualised artwork as a struggle between 'earth' and 'world'. A work of art, Heidegger argues in *The Origin of the Work of Art*, brings a world into being on the basis of 'earthly' elements, such as cloth and paint, bronze, vibrations of the air, etc. (Heidegger 1971). Experiencing a work of art is experiencing the coming into being of a meaningful world out of these material elements. Looking at Van Gogh's painting of a pair of peasant shoes in Heidegger's well-known example sets into motion a 'struggle' between the cloth and the pigment on the one hand, and the reality of the pair of shoes that arises out of these material elements on the other. In this struggle, there is no oppression and liberation, but interaction and mutual shaping. It would be strange to say that the paint oppresses the shoes or the other way round: the two are, necessarily, connected. Similarly, forms of human existence and arrangements of society are 'revealed' in the relations between technological materialities and the human beings who design, organise and use them.

But what can ethics still be, when the boundaries between humans and technologies disappear? In the symmetrical approach of Bruno Latour, where human and non-human entities play equal roles, and in the mediation approach of postphenomenology, where human practices and experiences are always technologically mediated, there no longer seems to be an 'outside' position with respect to technology. And if there is no longer an outside, from where could we criticise technology?

In order to articulate an alternative model for ethics it is helpful to connect to Foucault's approach to the phenomenon of 'critique' in his lecture 'What is Enlightenment?' (Foucault 1997a). In his analysis, Foucault is looking for an answer to what he calls 'the blackmail of the Enlightenment'. This blackmail consists of the fact that it is extremely hard to criticise the Enlightenment, as all attempts to do so are typically explained as being *against* it. Anyone who dares to open up a discussion about the Enlightenment raises the suspicion of being against rationality, democracy and scientific inquiry. Foucault, however, wants to explore if an *alternative* Enlightenment would be possible. This ambition is recognisable in the context of the ethics of technology. Blurring the boundaries between humans and technologies can easily be explained as giving up on ethics: because there is no longer a clear boundary to be defended, it might seem that 'anything goes'. Therefore, an alternative model for ethics needs to be developed.

As the title of his lecture suggests, Foucault was occupied primarily with the work of Immanuel Kant. In fact, he proposes an empirical and practical reinterpretation of Kant's ideas on the Enlightenment. Rather than aiming to transcend the empirical world, as Kant did, Foucault reinterprets Enlightenment as an *attitude* within the world. For Kant, as Foucault explains, Enlightenment was primarily a way out of 'immaturity', using 'reason' rather than accepting 'someone else's authority to lead us in areas where the use of reason is called for' (Foucault 1997a, p. 305). This way out of immaturity requires critique: critique can tell us under which conditions 'the use of reason is legitimate in order to determine what can be known, what must be done, and what may be hoped' (Foucault 1997a, p. 308). Critique, then, according to Foucault, must be understood as an attitude, an 'ethos'. It is the attitude of always looking for the limits of what seems to be given and self-evident.

Foucault calls this Enlightenment attitude a 'limit attitude'. This attitude is looking for 'The singular, the contingent, and the product of arbitrary constraints' in 'what is given to us as universal, necessary, obligatory' (Foucault 1997a, p. 315). Unlike Kant, though, Foucault does not want to transcend the empirical world into a transcendental realm. While Kant investigated the transcendental conditions of human reason, Foucault reinterprets critique—the 'enlightened' activity *par excellence*—as a form of practical self-inquiry. For Foucault, critique means investigating what has made us the beings that we are and what conditions have shaped our current form of existence. In order to be critical he does not *transcend* the limits of the empirical in order to find an 'outside' position, but rather positions himself *at* the limit. For Foucault, after all, there is no outside position from where to think. The human subject is always situated *within* the world to which it has a relation.

In the context of technology this means that the frameworks from which one can criticise technology are technologically mediated themselves. We can never step out of these mediations. The furthest we can get is to the limits of the situation we are in. Standing at the borders, recognising the technologically mediated character of our existence and our interpretations we can investigate the nature and the quality of these mediations: where do they come from, what do they do, could they be different?

As such, the Foucauldian limit attitude provides a 'way out' of the question of whether ethics of technology is still possible when we embrace the hybridising approaches of Ihde and Latour. Rather than letting ourselves be blackmailed by the Enlightenment—fearing that a non-modern conceptualisation of technology and society as interwoven would make it impossible to have a reasonable and normative discussion about technology—an alternatively enlightened 'limit approach' can offer a different, non-modern ethical approach to technology. It is not the *assessment* of technological developments from outside that is the central goal of ethical reflection then, but rather its *accompaniment*, 'from within', borrowing a concept from the Belgian philosopher Gilbert Hottois (Hottois 1996).

The crucial question in such a form of 'ethical technology accompaniment' is not so much where we have to draw a boundary between human beings on the one hand and technologies on the other. It is rather how we should give shape to the inter-relatedness between humans and technology, which has, in fact, always been a central characteristic of human existence. The limit attitude leads to an ethical approach that is not preoccupied with the question of whether a given technology is morally acceptable or not, but that is directed at improving the quality of our lives, as lived with technology.

Focusing on the intricate relations between human beings and technologies does not mean, to be sure, that all relations are equally desirable, and that rejection of a technology is no longer possible. Rather, it implies that ethics needs to engage more deeply with actual practices of design, use and implementation. Giving up an external position does not require us to give up all critical distance; it only makes sure that we do not overestimate the distance we can take. The Foucauldian limit attitude urges us to develop a 'critique' from within, engaging with how technological practices actually take shape and from a situation that is technologically mediated itself.

In line with Michel Foucault's ethical work (2010), this 'technology accompaniment' can be seen as a form of 'governance'. By deliberately shaping one's involvement with technology and with the effect technology can have on one's existence it becomes possible to give direction to one's technologically mediated subjectivity. Governance needs to be distinguished sharply from 'steering'. Governing technological developments implies a recognition of their own dynamics and of the relatively limited autonomy human beings have in their relation to technology. Human beings are 'implied' in technological developments, just as technologies are 'implied' in human existence. From this 'hybrid' point of view, in which humans and technologies are closely intertwined, the modernist ambition to 'steer' technology and to 'protect' humanity against technological invasions needs to be replaced with a more modest ambition to 'govern' the development of technology by taking its social implications into account, and to 'govern' one's subjectivity in relation to those technologies. By governing the relations between humanity and technology, we give up the idea that we can control technology; rather, we aim to understand how technology affects us and explicitly get involved in that process by critically designing and using technologies from the perspective of their mediating powers in human existence and our technological society.

Analysing the morality of technology

Governing technological developments requires us to be able to understand the impact of technologies on society. Here, the theory of technological mediation, which developed out of Don Ihde's post-phenomenological approach to technology, can be a helpful framework (Ihde 1990). Mediation theory approaches technologies as mediators of human-world relations. When used, technologies establish relations between human beings and their environment. These relations have a hermeneutic and an existential dimension: 'through' technologies, human beings are present in the world and the world is present for human beings. Technologies, in other words, help to shape human experiences and practices (cf. Verbeek 2005). Cell phones help to shape how human beings experience one another, while intelligent speed adaptation technologies help to shape people's driving behaviour in cars.

The central idea in mediation theory is that technologies do not simply create *connections* between users and their environment, but that they actively help to *constitute* them. Cell phones are not neutral intermediaries between human beings, but help to shape how humans are 'real' for each other. And, likewise, sonograms are not simply 'pictures' of a foetus, but help to shape what the unborn child is for its parents, and what these parents are in relation to their unborn. Mediation does not take place between pre-given entities, but helps to constitute the reality of these entities.

This mediating role of technologies has important implications for the ethics of technology. Mediation theory shows that human actions and decisions are fundamentally technologically mediated. In a 'material way', technologies help to give answers to the moral questions of 'how to act' and 'how to live'. Moral actions and moral decisions are not the product of autonomous human agency, but take shape in close interaction with technological artefacts. Technologies are morally charged, so to speak. They actively contribute to human morality (cf. Verbeek 2011).

Insights in technological mediation and the moral significance of technology can be a basis for an 'accompanying' ethics of technology. As I will explain later, mediation theory can help to make a (moral) 'mediation analysis' of a technology-in-design in order to inform the activities of designers.

Three levels can be distinguished at which mediation analysis can inform the work of designers. First, performing a mediation analysis can help them to *anticipate* the moral dimensions of the technology-in-design, for instance in order to avoid undesirable mediating effects. Second, mediation analysis can be the basis for *assessing* the quality of expected mediations. Making such assessments, to be sure, does not imply a shift back from 'accompanying' to 'assessing' technology; rather, it should be seen as a fully-fledged part of 'technology accompaniment'. And, third, mediations can be explicitly designed into a technology. In this case, we can speak of an explicit 'moralisation' of technology, following Dutch philosopher Hans Achterhuis (Achterhuis 1995).

Anticipating mediations

In order to anticipate the mediating roles of a technology-in-design it is important to try and make an analysis of the potential mediating roles the technology could have in the future. Making such an analysis is a complicated thing; it is never possible to make exact predictions about the future. Still, insights from mediation theory can guide the designer's imagination in order to make an 'educated guess'. Mediation theory then functions as a 'heuristic tool' to look for possible mediation effects.

In order to present such a heuristic tool here, I will integrate various elements of mediation theory that have been developed over the last few years into a coherent framework that can be used to anticipate technological mediations. These elements focus on the *locus*, the *type* and the *domain* of mediations: which 'point of application' does the technology have, which forms do its impacts have and which aspect of human existence does it affect?

Points of application

For conceptualising the locus of mediation we can connect to Steven Dorrestijn's work on behaviour-influencing technology. In his book *The Design of Our Own Lives* (Dorrestijn 2012) he distinguishes four points of application from which technologies can have an effect on human beings. Dorrestijn divides the space around human beings into four quadrants: 'to the hand', 'before the eye', 'behind the back' and 'above the head'.

Mediations 'to the hand' are physical: speed bumps that make it impossible to drive too fast or turnstiles that force metro users to buy a ticket. In addition to this, mediations 'before the eye' have a more cognitive nature. Technologies give cues or signals here in order to influence our behaviour: navigation systems that beep when one drives too fast or smart energy meters that persuade people to switch off the standby mode of their equipment. Mediations 'behind the back' concern the contextual and infrastructural role technologies can play. Technologies do not influence human behaviour and experiences here directly, but rather by shaping an environment in which specific forms of action and behaviour can come about. These influences can be located at a high level of abstraction, as demonstrated by Dorrestijn when discussing how the development of spectacles contributed indirectly to the success of the printing press because without spectacles a substantial part of the population would not be able to read. But they can also be seen at a more mundane level that might have more direct relevance for designers, such as the influence that a reliable and easily reachable public transport system will have on people's decision to take the car or the train to go to work. The fourth quadrant Dorrestijn distinguishes concerns the most abstract type of influence. Here, it is not *technologies* but 'technology' in general that influences the human being, with theories ranging from utopian optimism to dystopian pessimism. Because designers do not design 'technology' as such, but only concrete technologies, we will not take this quadrant into account.

In sum, three main points of application become visible here: technological mediations can be physical, cognitive or contextual. Technologies help to shape human-world relations through the physical-sensorial relation they create between humans and world, through the cognitive relation they create by giving information that can inform actions and decisions, and by creating a material and meaningful infrastructure that indirectly guides human actions and decisions.

Types of mediation

The second element of this heuristic tool for mediation analysis concerns the type of mediation that is involved. Earlier, I made a distinction between coercive, persuasive and seductive forms of mediation (Verbeek 2009). Some technologies actually force their users to behave in specific ways, like speed bumps or automatic speed limiters that require motorists to slow down. But not all mediation has this compelling form.

To stick to the example of car driving: an econometer in a car, which gives feedback about one's fuel consumption, does not force drivers to drive more economically, but rather persuades them to change their driving style. And the optical narrowing of roads—which is common in the Netherlands, where the central lane dividing line on roads is replaced with two lines at the sides of the road to create two biking lanes—seduces people into driving more slowly.

Design researcher Nynke Tromp, who has extensively studied the various types of influences that technologies can have on human beings, suggests categorising these influences along two dimensions (Tromp et al. 2011). One dimension indicates the force of the influence (weak versus strong) and the other its visibility (hidden versus explicit). This results, again, in four quadrants. While coercive influences are both explicit and strong, persuasive influences are explicit and weak: one can easily ignore the beep that suggests wearing a seat belt in a car, but one cannot avoid the effects of a speed bump. And while seductive influences are both weak and implicit, 'decisive' influences are implicit, but strong. Placing a coffee machine in the hall of a company will seduce people into having more informal interactions, while deliberately designing a multiple-story building without an elevator implicitly, but strongly, decides for people that they will have to use the stairs. In sum: technologies can force, persuade, seduce or decide for people.

Domains of mediation

The third, and final, element in a tool for mediation analysis is the domain of mediation. A first division of domains follows directly from the elementary framework of mediation theory: the existential versus the hermeneutical domain. Technologies help to shape how human beings are in their world and how the world can be there for human beings; they mediate actions and perceptions, practices and experiences. But besides the existential/hermeneutic distinction, we can also distinguish between individual and social mediations. Technologies do not only help to shape the practices and experiences of individuals, but also inform social practices and frameworks of interpretation. As Jantine Bouma has shown (Bouma et al. 2009), communication systems in co-housing communities have an important influence on social practices in these houses, and digital whiteboards have important implications for practices of teaching and learning in classrooms. Obviously, in all of these practices, there are also individual human–technology relations at work, but the eventual mediating effect reaches beyond this individual level. When designing a whiteboard, not only the effects on the individual experiences and actions of teachers and pupils are important, but also the effects on the learning process, the roles of teachers and pupils, etc.

The following table draws together all of these dimensions of mediation. Going through this table can help designers to stimulate their imagination and anticipate the possible mediating effects of the technology they are designing. Please note that the elements of mediation can be combined in all permutations. Physical, cognitive and contextual effects of technologies can have coercive, persuasive, seductive and decisive forms at the individual, as well as the social, level, and both in the hermeneutic and in the existential realm.

Locus	Form	Domain
Physical	Coercive	Individual: experience
Cognitive	Persuasive	Individual: actions
Contextual	Seductive	Social: frameworksof interpretation
	Decisive	Social: social practices

Assessing mediations

The second level at which designers can incorporate mediation in their work is the level of 'mediation assessment'. Here, mediations are not only anticipated, but also explicitly evaluated. To be sure: this activity of assessment should be seen as an element of the accompaniment of technology not as an alternative to it. Such 'accompanying evaluations' of technologies can take place by using an adapted version of the model that Berdichevsky and Neuenschwander developed for the evaluation of persuasive technologies (i.e. technologies that are designed to have explicit persuasive effects). Berdichevsky and Neuenschwander propose to evaluate persuasive technologies in terms of (i) the intentions of the designer; (ii) the methods of persuasion used; and (iii) the outcomes of the persuasion (Berdichevsky and Neuenschwander 1999). When we translate this to mediation theory, we can see persuasion as only one of many forms that mediation can take. Also, we can add a fourth dimension: besides intended mediation, there can also be implicit mediations, which occur without having been explicitly intended, but for which designers can feel partly responsible because they could have anticipated it.

To assess the quality of the mediations that are anticipated with the help of the mediation analysis tool described earlier, then, four steps become visible, which are inspired by Berdichevsky and Neuenschwander, but expand their framework to mediation theory (cf. Verbeek 2011, pp. 106–7).

The first step is rather obvious: if designers are working explicitly on a behaviour-influencing technology they could assess the *intended mediations* of the technology-in-design, i.e. the mediations that are deliberately designed into the technology. The central question here is: What arguments can be found in favour of and against these intended mediations, and the intentions behind them?

More interesting, though, is the assessment of the mediations that are *implicit* in the design. The heuristic tool for mediation analysis that was elaborated earlier can serve as a basis for this. It enables designers to anticipate unintended mediations that the introduction of the technology might bring about. Therefore, it also makes these mediations open for moral discussion: What arguments can be given to support or avoid these mediations?

A third element in assessing mediations concerns the *forms of mediation* involved. As indicated earlier, mediations can be strong or weak, and explicit or hidden. In specific circumstances, specific forms of mediation might be more desirable than others. For many, seducing car drivers to slow down in specific zones without them being explicitly aware of it will be less problematic than secretly seducing customers to buy much more than they actually intended by means of subliminal stimuli, such as emotion-evoking smells and colours.

Fourth, the eventual *outcomes* of the technological mediations—the actions and decisions that eventually take shape, as well as the social practices and frameworks of interpretation—can be assessed. All explicit and implicit mediations have effects, both at the individual and at the social level. These effects might be radically different from the original intentions of the designer. Speed bumps, for instance, will not only mediate the driving behaviour of motorists, but can also attract skateboarders, whose activities do not necessarily enhance road safety.

Designing mediations

The third, and last, level of the ethical accompaniment of design is the actual *design* of mediations 'into' technologies. Here, the role of designers becomes more socially invasive. Rather than checking for unwanted mediations, or explicitly assessing the implicit and explicit mediations involved in the design, they can also deliberately design for mediation. To which degree can this form of 'accompanying technology' be seen as morally desirable?

Moralising technology

An approach such as this was proposed in the 1990s by Dutch philosopher Hans Achterhuis, in his article 'De moralisering van de apparaten' ('The moralisation of devices'; Achterhuis 1995). Achterhuis argued that we need to end the constant moralising in environmental discourse. If all of the ideals of a small group of environmental activists were to be realised, even the smallest details of our existence would be subject to moral reflection, Achterhuis stated. The power of the lights we use in the house, the length of time we spend taking a shower, the fuel consumption related to our driving style—if everything becomes morally charged and subject to constant reflection, ordinary life will become impossible. Instead of moralising one another, Achterhuis states, we need to start moralising our technologies. We should delegate specific moral tasks and responsibilities to technologies, knowing that they are widely supported and that we are too weak and too limited to put all moral responsibilities on our own shoulders.

Achterhuis' approach was heavily criticised, especially because people were afraid that it would threaten human freedom. If we delegate morality to things, we seem to gamble with the crown jewel of humanity: our capacity to make autonomous decisions and to take moral responsibility for our actions. I do not share this criticism, as might be clear from the first section of this chapter. When taking the phenomenon of mediation seriously, all human actions are technologically mediated. The explicit design of mediating technologies is not immoral from this point of view, but rather refusing to take responsibility for these mediations.

Libertarian paternalism

Precisely at this tension between being steered and maintaining autonomy, Richard Thaler and Cass Sunstein seem to have found an answer. In their book *Nudge* (Thaler and Sunstein 2008), they make a case for designing our material surroundings in such a way that it influences us in a positive sense without taking control away from us. A nudge is a tiny push, a small stimulus that guides people's behaviour in a certain direction. Our material world is full of such nudges, Thaler and Sunstein claim, varying from photocopying machines with a default setting of single-sided copies to urinals with a built-in image of a fly to seduce men to aim for it. Thaler and Sunstein propose that we design these nudges in an optimal manner, so that we can guide our own behaviour in directions that are widely considered to be beneficial.

The central idea in their approach is that human decisions are, to a considerable extent, organised and prestructured by our material surroundings. When we make choices two systems are at work in our brains, which Thaler and Sunstein call an 'automatic system' and a 'reflexive system'. Most of our decisions are made automatically, without explicit reflection. But, for some decisions, we really have to stop and think: they require reflection and critical distance.

To a significant degree, our automatic system is organised by our material surroundings. To use one of Thaler and Sunstein's examples: when fried snacks are within reaching distance in a company's canteen and the salads are hidden behind refrigerator doors, it is very likely that many people will choose the less healthy food. The layout of canteens gives nudges in a certain direction. If we want to take responsibility for such situations, we must learn to think critically about nudges. If we can design them better, we, in fact, design our automatic system to behave in a more desirable way. Thaler and Sunstein therefore call such design activities 'choice architecture': the design of choice situations. We need to rewrite the default settings of our material world.

But these activities of choice architecture should never close down the reflexive system. For Thaler and Sunstein, it is extremely important that nudges always remain open to reflection and discussion, and can move from the automatic to the reflexive system. This is why they indicate their approach as 'libertarian paternalism'. It is paternalistic because it explicitly exposes people to nudges in a direction that is considered desirable. But it is also libertarian because these nudges can always be ignored or undone in all freedom. Just like everyone is currently free to use both sides of the paper when copying, even though the standard setting is one side, no one should be forced to eat a salad and pass up the croquettes in a 're-nudged' cafeteria.

Thaler and Sunstein's way out of the dilemma between influencing behaviour and respecting autonomy, therefore, is the 'opt-out': by drawing on our reflexive system, we should always be able to move away from the nudges. Every act of paternalism is compensated by the explicit possibility to take a libertarian stance towards it.

The question, however, is whether this libertarian–paternalistic attempt to tame the morality of things is a real solution. Both the libertarian and the paternalistic elements of Thaler and Sunstein's approach take a separation of humans and technologies as a starting point, with a libertarian focus on saving human autonomy. On the one hand, nudges are the result of paternalistic human design, while, on the other hand, the people subjected to this paternalism always have the libertarian possibility of ignoring it. The phenomenon of technological mediation has no place whatsoever in Thaler and Sunstein's approach: nudges are discussed as instrumental interventions of paternalistic designers that can be either accepted or rejected by critical users. The boundaries between humans and technologies remain fully intact here; rather than critically *interacting* with technology to shape one's existence, the primary form of criticism is to *opt out*.

Design as material ethics

If libertarian paternalism is no real option, then what could the alternative be? Do the boundary-blurring approaches of actor–network theory and postphenomenology simply urge us to accept paternalism? Should designers be allowed to steer our actions and decisions behind our backs, without giving us the possibility to reject these interventions in our lives? The answer to this question is both yes and no. Yes, because contemporary approaches in philosophy of technology show that human actions and decisions are *always* technologically mediated. There is no way in which designers could avoid having an effect on human existence. But, at the same time, the answer is no because this effect is not necessarily exerted behind our backs.

Once we see the phenomenon of technological mediation, we can always develop a critical relation to it—in the Foucauldian sense of critique. Not to step out of the field of mediations, but to stand at the boundaries of that field in order to find out which forces are exerted upon us, and how we can shape our own lives in interaction with these forces. While we can no longer conceive of ourselves as autonomous beings because of the fundamentally mediated character of our lives, we can still develop a free relation to these mediations. Without being able to undo or ignore all of them, we can critically and creatively adopt them. Being a citizen in a technological society requires a form of 'technological literacy'. Not in the sense that every citizen needs to understand all the technical details of the devices around them, but in the sense that we develop a critical awareness of what technologies do in society.

This technological literacy on the part of technology users, then, is a necessary complement to the responsible ways of dealing with mediations that we can ask of designers. While designers need to take responsibility for the mediating roles of their products by anticipating, assessing and designing the implicit and explicit mediations that are involved, users need to take responsibility for their own, technologically mediated existence.

Conclusion: An example

Let me conclude by giving an example of such a 'moralising technology'. It is a telecare technology for patients with chronic obstructive pulmonary disease (COPD)—a lung disease that dramatically reduces one's lung capacity and which is potentially lethal. Patients with this disease need to have their lung capacity checked on a regular basis and need to continuously adapt their activity pattern very carefully to the situation of their lungs and their physical condition. The problem for COPD patients is to find the right balance between training their lungs enough to slow down the progression of the disease on the one hand and not demanding too much of themselves on the other. The 'COPD.com' system aims to help patients to find this balance. COPD.com is primarily a 'disease management system'. Its basis is a so-called 'Body Area Network' that integrates various sensors to monitor a patient's activity level and physical condition. The data generated by this network are translated into a coaching programme that is accessible via a web portal. Patients can log in, find information about their condition and get advice about the optimal exercise they should have.

This system is quite invasive: it monitors a patient's activities in a detailed way and advises them about what to do. If ethical reflection were to limit itself to assessing whether this technology is morally desirable or not, the only relevant questions would be whether this invasiveness stays within acceptable norms, and whether the system is safe and reliable. From the point of view of mediation theory, though, the most interesting questions relate to the effect of the system on the daily lives of patients. A life with COPD takes on a new shape when COPD.com starts to play a role in it. Using the heuristic tool described above can result in various anticipated mediations. For instance, what does the continuous monitoring of one's activity level and one's physical condition do to the self-understanding of patients? The system could result in a far-reaching medicalisation of people's lives. Also, responsibilities will start shifting. Rather than nurses and doctors, patients themselves now become central agents in the treatment of the disease. On the one hand, this enhances the autonomy of the patient, but, on the other hand, it also makes patients more responsible when things go wrong. And, the interaction between care giver and patient will change.

Part of the work of the nurse will shift from having conversations with patients about their condition to making the system work optimally so that the system can have these 'conversations', for instance.

These are only a few examples of what a mediation analysis could reveal when designing a technology like this. An obvious next step, after assessing the quality of these mediations, would be to incorporate the results of this analysis in the design process itself. Designers could choose to make it impossible for patients to monitor their physical condition more than twice a day, for instance, in order to prevent an unwanted medicalisation occurring. Also, protocols for using COPD.com could require a regular visit to a nurse or doctor to ensure that all relevant aspects of the life of the patient are known, to keep up an interpersonal relation of care, and to see how the interaction between the patient and the system takes shape.

COPD.com is definitely a 'moralising technology' in the sense that it actively prescribes its users how to behave. Patients who use the system, though, are not reduced to slaves of the machine. They can be helped by nurses and doctors to develop a critical relation to the system: to evaluate the quality of the advice it gives, to take the liberty to ignore or modify the advice the system gives, and to find a good way of dealing with the new lifestyle and self-image that the system introduces.

In a very modest and minimal way this example shows how much ethical space there actually is on the laboratory floor. Rather than assessing this technology from an external perspective, by focusing on the question of whether it should be considered a desirable technology or not, the proposed method for 'accompanying technology' makes it possible to get involved in the practices of design and use that surround this technology. First, by making a mediation analysis of COPD.com the ways in which this technology helps to shape the activities and experiences of COPD patients and the people in their environment become visible. Where does it exert its influences: is it physical, cognitive or contextual? Are the influences coercive, seductive, persuasive or decisive? And do they take place at the individual level or at the social level? Second, this analysis can be the basis of a careful assessment of the quality of these various mediations: What are their effects and how can these be valued? On the basis of this a third step can be made: How to design desirable effects into the technology? How invasive can a technology be without being experienced as an obstacle rather than an aid? How to design successful mediations?

Taking seriously the idea that technology and society continuously help to shape each other therefore does not imply the end of ethics, but rather a new beginning. The design of technologies itself has become an intrinsically moral activity. Responsible design requires the anticipation, assessment and explicit design of the mediations that the technology will introduce in society. Designing mediations is, inevitably, a jump into an unknown future, and will always have an experimental character. But, by systematically anticipating and assessing the mediations involved in the design, we at least organise these experiments as responsibly as we can. Ethics on the laboratory floor does not only involve processes of scientific innovation, but also of technological design.

In order to give such an ethics of 'technology accompaniment' a firm basis, two strategies can be followed. On the one hand, further research needs to be done into the specificities of this accompanying form of ethics. By using empirical methods from sociology and cultural anthropology it becomes possible to investigate in which ways users and designers make technologies morally significant in their practices and conversations. Such research can shed more light on the moral significance of technologies-in-design. On the other hand, ethical reflection should be brought to the field of technology design. Design methods need to be expanded with tools for the anticipation, assessment and design of mediations. And, most importantly, design schools will need to teach their students that design is an intrinsically social activity in which designers should learn to take responsibility for the ways in which they intervene in society.

Notes

1. This section incorporates reworked fragments from my article 'Resistance is Futile', forthcoming in *Technè* 2013.
2. This section incorporates reworked fragments from my article Verbeek, P.P. (2012) 'Politics at Issue. On Art and the Democratization of Things', in *Open 24: The Politics of Things – What Art & Design do in Democracy*, pp. 18–29 (Rotterdam: NAI Publishers).
3. COPD.com is developed by Roessingh Research and Development, the University of Twente, and Medisch Spectrum Twente; for more information visit http://www.copddotcom.nl.

References

Achterhuis, H. (1995) 'De moralisering van de apparaten', *Socialisme en Democratie*, 52(1): 3–12.

Berdichewsky, D. and Neuenschwander, E. (1999) 'Toward an Ethics of Persuasive Technology', *Communications of the ACM*, 42(5): 51–8.

Bouma, J.T., Voorbij, A. I. M., and Poelman, W. A. (2009) 'The Influence of Changes in the Physical and Technical Design on Social Interactions in a

Cohousing Community', in Durmisevic, E. (ed.) *Lifecycle Design of Buildings, Systems and Materials* (Enschede: International Council for Building Research Studies and Documentation (CIB) and the University of Twente).

Dorrestijn, S. (2012) 'The Design of our Own Lives', dissertation (University of Twente).

Foucault, M. (1997a) 'What is Enlightenment?', in Rabinow, P. (ed.) *M. Foucault, Ethics: Subjectivity and Truth* (New York: The New Press).

Foucault, M. (2010) *The Government of Self and Others: Lectures at the Collège de France 1982–1983*, Davidson, A. I. (ed.) translated by G. Burchell (New York: Palgrave Macmillan).

Heidegger, M. (1971) 'The Origin of the Work of Art', in Anderson, J.M. and Freund, E. H. (eds) *Poetry, Language, Thought* (New York: Harper & Row).

Hottois, G. (1996) *Entre symboles et technosciences. Un itinéraire philosophique.* (Paris : Editions Champ Vallon).

Ihde, D. (1990) *Technology and the Lifeworld* (Bloomington/Minneapolis: Indiana University Press).

Latour, B. (1993) *We Have Never Been Modern* (Cambridge, MA: Harvard University Press).

Thaler, R. and Sunstein, C. (2008) *Nudge: Improving Decisions About Health, Wealth, and Happiness* (New Haven: Yale University Press).

Tromp, N., Hekkert, P., and Verbeek, P. P. (2011) 'Design for Socially Responsible Behavior: A Classification of Influence Based on Intended User Experience', *Design Issues* 27(3): 3–19.

Verbeek, P.P. (2005) *What Things Do: Philosophical Reflections on Technology, Agency, and Design* (University Park, PA: Pennsylvania State University Press).

Verbeek, P.P. (2009) 'The Moral Relevance of Technological Artifacts', in Duwell, M. (ed.) *Evaluating New Technologies* (Dordrecht: Springer).

Verbeek, P.P. (2010) 'Accompanying Technology: Philosophy of Technology after the Ethical Turn', *Techné: Research in Philosophy and Technology* 14(1): 49–54.

Verbeek, P.P. (2011) *Moralizing Technology: Understanding and Designing the Morality of Things* (Chicago/London: University of Chicago Press).

Part II
Case Studies

5

Co-shaping the Life Story of a Technology: From Technological Ancestry to Visions of the Future

Simone van der Burg

Introduction

The timing of any ethics on the laboratory floor poses a challenge, which has been famously described by Collingridge's dilemma of control. This dilemma claims, on the one hand, that it is hard to control a technology once it has finished developing because at that point too many parties—researchers, producers, investors—have an interest in putting it on the market. Anyone who wants to prevent this from happening, who wants to impose restrictions on its use or change the technology itself, will have to come with heavily weighed arguments—concerning life and death—in order to reach her/his goal. Any attempt to control technologies at an earlier stage of development, however, does not have a chance of being successful either. According to the second horn of Collingridge's dilemma, efforts to govern technology during research or development cannot be productive because at that stage it is still uncertain whether these technologies will be realised at all. It is therefore unclear what could be problematic about them and difficult to determine how they need to be governed (Collingridge 1980).

The second horn of the Collingridge dilemma implies that technologies that are being researched are not yet 'something' because there is not yet a device about whose characteristics we can speak (Johnson 2007). It is therefore impossible to speak meaningfully about their morality. This relates to the very common intuition—expressed philosophically in Bertrand Russell's extensionalist theory (1905)—that there needs to be 'something' before we can ascribe predicates to it, including moral predicates, such as 'good', 'bad', 'just', 'generous', 'honest', 'respectful', 'democratic', etc. In the extensionalist theory, sentences acquire meaning when they are about something. The subject identifies the topic of the proposition; the predicate says something about it. Once the subject has been found, the proposition is ready to convey its information about it. But if the subject is lacking, or there is uncertainty about whether it exists or will exist, it seems a waste of time to say something about it.

This view of technology as a 'subject' to which predicates can be ascribed is nowadays often replaced by a more dynamic picture. Authors in science and technology studies, as well as in philosophy of technology, abandon describing technologies as 'objects' with 'features', 'characteristics' or 'functions', and instead identify what these technologies are in relation to what they allow people to do, to experience or to be. Bruno Latour's descriptions, for example, draw attention to how technology changes how human beings act and interact, what motivations they develop for action, what skills and habits they develop, and what goals they strive to reach; Don Ihde's work concentrates on how technologies alter the human experience of the world around them, of other people and themselves (Ihde 1990; Latour 1992, 1993). Building on the work of Latour and Ihde, Peter-Paul Verbeek developed his own philosophy of technological mediation in *What Things Do* (2005), in which he analysed how understandings of human life, including understandings of the good life, always generate in contexts in which artefacts are also present. None of these authors adopt an objectifying perspective to technologies, in which sentences are articulated about technologies and are out to verify or falsify them; rather, they perceive technologies as part of contexts of interaction to which they offer an active contribution.

These inspiring dynamic perspectives on technologies, however, continue to give prevalence to technologies that have already finished developing; they rarely pay attention to the ways in which technologies in earlier phases of development differ from technologies in use. While Latour describes a technology that is being designed in *Aramis* and Verbeek developed an ethics of design in his book *Moralizing Technology* (2011), as well as in his contribution to this volume, neither of these authors thematises the distinction between actual and imagined mediations. Technologies that are being designed and technologies already in use are referred to as implying the same dynamic exchange with people. The difference is that when the technology is being designed, the interaction is imagined and it is still uncertain whether it will become reality. Technologies during design are therefore a kind of 'prospective presences' or 'actualities'; they are imaginative copies of future human–technology interactions, which may eventually materialise.

In this chapter I will discuss a case study that stems from my own work as a moral philosopher in the Biomedical Photonic Imaging Group at the University of Twente (the Netherlands). In this case study I will attempt to bring forward ways in which the specific technology during research differs from technologies that are being used. Futhermore, in supplement to the future-oriented approaches that are included in the contributions to this volume by Peter-Paul Verbeek, but also in the ones by Marianne Boenink and Armin Grunwald, I will plead for a study of the ancestry of the technologies that are being researched, as this draws attention to characteristics that technologies may come to embody, and which they may come to repeat when used, in the future.

Photoacoustics and acousto-optics: A case study

The project in which I was engaged as an ethicist was funded by NWO (Netherlands Scientific Research) and was formed in cooperation with STW (Foundation Technology Research), which is a subdivision of NWO charged with the funding of research into technology. The end point of research funded by STW is the production of a prototype, which has to be tested on users in subsequent phases of research. STW does not fund what it calls 'fundamental' research, but research that is done before the industry is interested in financing its further development. STW, however, does want to stimulate the researchers to build a relationship with the industry. It therefore requires scientists to involve members of the industry in their so-called 'users committee', which also contains future users, such as—in the case of the technologies I was studying—physicians. This committee of 'users' is thought to facilitate the further development of the technology, as well as its marketability. During STW-funded research, members of the industry do not yet have to provide substantive funding in exchange for their involvement; it is enough if they deliver services in kind, such as material or man hours. The principal goal of the involvement of the industry is that scientists receive feedback on their research from a more market-oriented perspective. If the research leads to results that can be patented, however, the industry that was part of the users committee gets the first chance to acquire it.

STW also required the scientists to cooperate with my research. The first group with which I was involved worked on an acousto-optic device, which was researched for its capacity to non-invasively monitor chemical substances in blood, such as oxygen, but, in the future, also glucose, cholesterol and lactate. Later on, I also became involved in a second research team that researched photoacoustic mammography, which is an innovative device for non-invasive detection of breast cancer. Both technologies use sound, as well as light, although they aim for different applications. They therefore belong to the same technological 'family'.

In order to come to an understanding of what these technologies are, and how they matter morally, I engaged, essentially, in three activities: (i) I looked at experimental set-ups in the laboratory, (ii) studied the theoretical and experiential background of these set-ups (their 'past'), and (iii) interviewed the scientists to find out what futures they envisioned for these technologies. A study of the 'past' of the technology flowed naturally from visits to the laboratory. In the laboratory it is possible to 'see' ethical aspects of the technology that flow from its history in other contexts of research and use. In a laboratory experiment the technology lies bare and it is therefore possible to acquire a specific 'inside' perspective into the characteristic functionings that the technology may reveal in later stages of its development. But there is not yet one 'thing', one 'device' or 'artefact'. In fact, a combination of artefacts was visible in the laboratory experiments in the specific researches in which I was involved.

In the case of the acousto-optic monitoring device, for example, the experimental set-up included a laser, an acoustic transducer (ultrasound), a computer and a substitute for the human body called, interestingly, a 'phantom', which was a white pudding made of an Intralipid solution that had been frozen and defrosted several times in order to scatter light in a similar way to human tissue, and a tube with ink inserted into it that represented the blood vessel. Scientists assisted me in understanding what I was looking at and explained what they were testing. They already knew that infrared laser light is absorbed by things with colour, such as the ink in the tube, or the blood vessel. The degree of absorption of light by blood is indicative of the oxygen level in the blood: blood with little oxygen is dark red and will absorb more light than light-red blood, which contains a lot of oxygen. Vessels are, however, surrounded by tissue which scatters the light. The scientists were therefore researching whether they could distinguish the parts of the light (photons) that go through the vessel from the other light particles that are scattered by the tissue surrounding the vessel. They were doing this by means of ultrasound which was pointed directly at the vessel, as sound changes the movement of the photons (the pathway), enabling scientists to detect and produce an image of this movement. A visual display of the altered pathway of the light was shown on a computer screen in a graphic representation of the way in which the light moved: a wave. Photons that are modulated by ultrasound show a larger amplitude than photons that are unaffected by ultrasound.

The scientists explained that they were researching whether, by means of ultrasound, they are able to distinguish only those photons on which they have to concentrate to make the measurement of the oxygen level. And they added that, in the future, other chemical substances of blood could also be measured, such as glucose, cholesterol or lactate, using other colours of light. As a supplement to their explanations they provided articles that gave insight into previous research on which they had built upon. On the basis of what I saw in the laboratory, and what the scientists told me and gave me to read, I shaped my own literature search, which included a study on a technology that was an ancestor of the technology that they were researching, and on which technological principle they were building: the pulse oximeter. It is this technology that is the most important 'ancestor' of both acousto-optics and photoacoustics, and it is particularly interesting because it has already been used for more than 30 years in hospitals all over the world. I was therefore especially interested in the results of the tests on patients. This study of the 'past' of photoacoustics and acousto-optics—or, better, its ancestry—proved to be useful in discerning part of the moral characteristics of these technologies.

Past ancestry

The pulse oximeter is an optical instrument for the non-invasive monitoring of oxygen in blood and has been used widely in hospitals since 1981. The pulse oximeter is the little clip put on patients' fingers when they are to undergo surgery. It is an optical technology, as the measurement depends on the degree of absorption of a near-infrared light beam by blood, which, in turn, depends on the colour of that blood. Research into the acousto-optic monitoring of blood builds on this technology, but adds ultrasound with the purpose of improving its results by making it possible to focus solely on the photons that transgress the vessel and avoid the other photons that are scattered by tissue. In this way scientists are trying to improve on the imprecision of the measurement that the pulse oximeter provides because it focuses on all light and not only on the photons that go through the vessel.

Photoacoustic mammography also builds on the technological principle of the pulse oximeter. But it uses sound and light in a different way, and with a different purpose. Photoacoustic mammography attempts to depict the growth of blood vessels in the breast with an ultrasound image. The presence of extra vessel growth indicates that there is a tumour in the breast: because tumours need blood to be able to grow and spread there is always excessive vessel growth around them. Because it combines sound and light in such an ingenious way, photoacoustics could be called a *synesthetic technology*—synaesthesia being a neurological condition in which stimulation of one sense leads to experiences in another sense altogether. Photoacoustic mammography likewise acquires its function by means of a 'translation' of the perception of one sense into another. The light beam (associated with sight) is absorbed by blood vessels (just as in the case of the pulse oximeter). This absorption by blood will produce a temperature rise (tactile sense), followed by an expansion. Because the light beam is pulsed—it goes on and off—it will expand and shrink alternately, thereby producing a sound (auditory sense). This sound is visualised in an ultrasound image (sight) depicting the blood vessels that are present, which is the product that this technology aims to produce. The presence of extra vessel growth in the image is an indication that the detected lump in the breast is cancer and not a harmless cyst. An ultrasound image of the surplus vessels in the breast is therefore clinically relevant.

Both the acousto-optic monitor and the photoacoustic mammography build on the strengths of the pulse oximeter, but could also inherit its flaws. Since the mid-1990s, patient studies have indicated that the pulse oximeter gives different results on white skin than on dark skin, especially when oxygen levels are low (Van der Burg 2010). These differing results raise questions for new technologies that build on this technological ancestry concerning the possible repetition of this poor performance on dark skin. The trouble is that such poor performance is hard to notice in use, especially because users are unaware of the ways in which the pulse oximeter works (this is black-boxed), and, consequently, are unaware that skin colour could interfere with its performance. This was also the case with the pulse oximeter. After its introduction into the hospital, healthcare professionals were unaware that they had to interpret the results differently depending on the skin colour of the patient. It took 20 years to uncover the difference in performance on dark and light skin tones in patient tests. After that, new brands of the pulse oximeter were developed and introduced that make it possible to calibrate the performance of the technology to the skin tone of the patient.

This history of the pulse oximeter does, however, raise questions for new technologies, such as acousto-optics and photoacoustics, which build on the same technological principle. It is not clear, for example, whether an adjustment of the results according to skin colour is an option for the acousto-optic instrument. As it bases its measurement only on a selection of the photons that transgress the vessel, it raises the question as to whether its calibration is an option. As a lot of the light will be absorbed at skin level, and the measurement is based on only a few photons, research on whether it is possible to use it on dark-skinned patients at all needs to be performed. Based on the laboratory experiments, it seems quite improbable that this is an option for the acousto-optic technology for the monitoring of chemical substances in blood.

Similarly, research on whether the resolution of the image that photoacoustic mammography is able to produce on dark-skinned patients is clear enough to detect cancer, especially at an early stage of development, needs to be carried out. As the whole image that is produced depends on the absorption of light by blood, enough light has to be able to enter the body in order to be able to see anything at all. In dark-skinned patients less light will succeed in entering the body, which will affect the resolution of the image. If there are only a few extra vessels to absorb the light—or if these extra vessels are still very small, as is the case in early stages of cancer—it is questionable whether enough light will be absorbed to produce an image. This is, therefore, a question that needs to be taken into account in future phases of research when the technology is going to be tested on patients. It demands that dark-skinned patients have to be involved in these tests and that the performance on these patients has to be compared with the performance on light-skinned patients.

The performance of a technological ancestor of the acousto-optic monitor and the photoacoustic mammography on dark-skinned patients has, in this case study, led to moral concerns that are relevant to research. These concerns were well received by the scientists, who were unaware that dark skin posed a problem to the pulse oximeter in the first 20 years of use. Their carefully structured step-by-step approach to their research—without which it would be impossible to progress—forces them to order topics and questions in stages and phases, and considerations about patients turn up rather late in that ordering. During laboratory experiments scientists do not usually read articles about the performance of related technologies on patients; they just concentrate on technological issues and problems that they encounter in the laboratory. When the laboratory experiments are finished, however, and patient tests are being planned, they usually do not read these articles about patient tests of related technologies either because scientists take it that their new technology has developed away from these technological predecessors and will perform quite differently on patients. Consequently, a new technology may inherit the problems of its technological ancestors, while scientists are quite unaware of it. And, consequently, new technologies may enter the clinic, and it may take another few decades to detect their limitations and adapt them to the variable skin colours of patients, or to find out that they cannot be appropriately adapted.

A genealogy of a new technology, which reveals the performance of the technological ancestors of a new technology, could offer valuable information that helps to think in advance about problems that could occur in the future. These problems relate to the 'something' that the technology already is during research. Acousto-optics, for example, is not yet a device during research. Neither is it nothing; in fact, the experimental set-up tests and, in that way, reveals its actions before there is a device that actually 'does' them. In the case of this particular technology a characteristic way of acting is being researched before it has a 'subject' to which these actions can be ascribed. And, in parallel, features of the technology can be discerned while there is not yet a carrier of those features. Moral philosophers, however, are interested in the features of a technology and not primarily in their carrier. It therefore seems worthwhile to look at this 'something' whose features are there to be experienced in the laboratory, and not wait until they become black-boxed in a device and, consequently, difficult to distinguish.

These features, however, can also be studied prospectively, and I will report on that in the next section.

Visions of the future

On the basis of the scientist's stories about what they hope to create, I have attempted to come to a better understanding of what it is that they are creating (Van der Burg 2009). Marianne Boenink has termed this an act of *specifying* in this volume. If the technology is not yet there, and its actual functioning cannot be perceived yet, an exploration of the stories that scientists tell about it may be the only way to find out what it will become. Different members of the same research team may, however, articulate differing futures for the technology that they are researching. And sometimes they also disagree about what future is most realistic or which one will most probably materialise. These disagreements reveal instabilities and tensions in the identity of the technology as it is perceived now, but may also form the background of disagreements within the research team about the design of subsequent research steps or follow-up projects. The act of specifying, therefore, not only helps to make clearer what the 'something' is that is being researched, but also helps to distinguish in a more precise way in what directions the research may evolve further, which opens up the possibility of reflecting on the question of which one is preferable or more worthwhile pursuing.

In order to fuel these reflections I first articulated the visions of the scientists in scenarios. All of these scenarios focused mainly on the technological possibilities of photoacoustic mammography and the ways it could transcend the technologies that are used at present. Some focused on *screening*, others on *diagnosis*. If photoacoustics were to get a diagnostic function, for example, it would replace the puncture that is currently undertaken to diagnose breast cancer. A puncture is a procedure in which a pin 'shoots' into the breast and removes a little bit of tissue, which is subsequently tested in a laboratory. This puncture—in the current regular procedure—is taken when an x-ray mammography and an ultrasound indicate that there is a suspicious lump in the breast. Photoacoustic mammography could provide a non-invasive diagnosis, sparing women who have a non-malignant lump the puncture. Furthermore, photoacoustic mammography could provide more specific information that is relevant to the diagnosis: the oxygen level of the blood surrounding tumours indicates the speed with which cancer grows. A low oxygen level indicates a fast-growing tumour (which needs a lot of oxygen); a high level of oxygen is a sign of a slow-growing tumour.

If photoacoustic mammography is not used for diagnosis, but for screening, it would make x-ray mammography redundant. This would mean that the sensitivity of the technology needs to be explored further so that it can be trusted to detect cancer, even if the tumour is still small. According to the scientists it would be attractive to substitute current x-ray mammograms because x-ray uses harmful radiation and because it demands the breast to be made very flat, which causes women pain. If photoacoustic mammography were to be used to screen women it would liberate them from this harmful radiation and from the painful procedure of taking x-ray mammograms. Furthermore, one of the visions of the future of photoacoustic mammography, which needs to be researched further, is that it transcends the limits of x-ray mammography because it is able to produce a better image of the inside of young women's breasts. X-ray mammography misses 25% of breast cancer cases in young women because it has difficulty transgressing the glands in their breasts. In post-menopausal women these glands diminish, which is an important reason for including only older women in screening programmes. Scientists expect that photoacoustic mammography will be able to image the inside of young women's breasts and therefore also increases the chances of detecting cancer earlier.

These visions of the future, which arose from scientists' talk about the technology, were not yet real, as further research had yet to point out if photoacoustic mammography had the capacities that they hypothesised. But they are not speculative either. The visions of the future that I articulated and distinguished in scenarios are more like products of an 'educated imagination' that is constrained by knowledge about scientific methodologies and an erudite acquaintance with research on similar technologies that are researched elsewhere, and technologies that are already being used. It is this educated imagination that I tried to broaden a little in order to get a clearer insight into how the imagined capacities would change the lives of users in a clinical context, and to open a debate with the scientists about the desirability of that change.

My own qualitative interviews with ten patients and three clinicians pointed out that they much preferred the screening scenario. Both groups tended to not want to skip the puncture. Clinicians considered the puncture to be an unavoidable gold standard for a cancer diagnosis; patients stated they wanted to keep the puncture because they would never have believed they had cancer if nobody had actually investigated a biopsy taken from their bodies. Even if photoacoustic mammography was to become able to provide a more specific diagnosis—as one of the scenarios claims—one that also provides information about the speed with which the cancer grows, clinicians and patients would prefer not to have it. Clinicians said they did not have treatment specifically aimed at quick- or slow-growing tumours, and patients remarked that it would be extra difficult to deal with the diagnosis of a tumour that grows and spreads very quickly.

The screening scenario was more attractive to patients and clinicians, as photoacoustic mammography could take away an important part of the drawbacks of screening with x-ray mammography, such as its painful procedure (having to make the breast flat) and the use of harmful radiation. The avoidance of these drawbacks was considered a huge advantage of photoacoustic mammography by patients. Furthermore, patients valued that photoacoustic mammography would be able to provide more specific information about the lump in the breast, allowing a clearer distinction to be made right away between malignant tumours (with extra vessel growth around them) and non-malignant tumours that need no further intervention. Clinicians, however, mentioned that they are accustomed to looking for tumours and that it demands a switch in their focus to look for vessel growth; they prefer to see the tumour as well, and prefer to use ultrasound or x-ray mammography in combination with photoacoustics. Women also thought it an advantage that young women could be screened, although they also mentioned a risk of medicalisation if the age group of women who receive screening is enlarged. The capacity of photoacoustic mammography to screen young women also takes away a technological argument to limit the population that is invited for screening to older women (post-menopause), as is presently the case in the Netherlands. Patients appreciated the relevance of the question of whether it is desirable to enlarge the screening practice to include young women as well, which raises issues of justice and medicalisation of the healthy period of life (Van der Burg 2009).

These scenarios, and the opinions of patients and clinicians, gave rise to themes for ethical consideration, such as the reliability of technologies and the trust that clinicians and patients are able to place in them, the issues of risk, harm and pain, and of security in the future. But also of justice, medicalisation of healthy life, and growing costs related to the enlargement of screening practices, which raises issues about the just allocation of scarce resources. All these issues raised interesting conversations with the scientists. Generally, they acknowledged their relevance and importance, and eagerly added to the reflection about them. However, they did not appropriate all of them as being *their responsibility* as scientists who give shape to a new technology. As scientists they easily included the skin colour issue into their perception of their responsibility (interestingly, they appropriated it so well that they claimed it was a *technological problem and not an ethical one*), and they recognised that a focus on screening would be a more desirable priority for envisioned users. Screening, however, did not eventually become their main focus afterwards. Funding opportunities in the area of nanotechnology steered their research activities in the direction of the further development of the diagnostic function in order to enable it to detect cancer at the nanoscale.

Future orientation and past ancestry

It is often mentioned that it is a challenge to do ethics on the laboratory floor because it is not yet established what the technology is going to be. What I find interesting in this case study, however, is that it shows that while the technology is not yet there during research, its characteristics—which may be capacities, actions or features—are already present in several ways. The laboratory is a unique space in the sense that it allows any spectator to look inside and see how technology 'works', even if there's not yet a device present to which such 'workings' could be ascribed. In an experimental set-up, the technology does not, of course, have its definitive capacities—but it is at least possible to see and understand the ones that are being investigated, and ask questions about them with the purpose of enhancing ethical reflection. This will become more difficult at a later stage in the development of a technology when the functionality becomes confined to the secret interior of a usable device.

The openness of the technology in a laboratory experiment allows questions to be asked about its composites and history, as well as about its future. The history may allow coming to grips with values, which the technology may embody without the scientists being aware of it. But the capacities that are revealed in the test set-up may also allow more specific questions to be asked about the scientists' visions of the future, which also relates to their planning of future research. While the 'something' they are trying to create does not yet exist, the capacities that they are researching do exist in some form and can become visible in the laboratory.

The proposal here is therefore to focus on the capacities of technologies 'in the making'. The ethical significance of technologies 'in the making' may be interpreted on the basis of the capacities that are envisioned for them in the future, and the capacities that they may inherit from predecessors. Borrowing the metaphor 'biography' of Lorraine Daston (2000), the proposal here is that ethicists who are engaged in the laboratory write a biography of the technology that is being researched, trace its ancestry, and describe the characteristics that it may inherit and give to future generations. A biography of the technology gives insight into its remembering, a word taken in the literal sense, that is *re-membering*, putting together the members, parts, limbs *again*. As the research process into some technologies (but not all) sometimes involves a re-assembling of parts, a study of the past of its members is an enterprise that can shed light on the character of the newly developing technology, including the features it will play out in future use.

The metaphor of a biography suggests an analogy between the development of a technology and a human life. The life of a technology depends, just like the life of a human being, on inherited capacities, but also on an orientation towards the future. It is towards this future that some capacities may come to bloom, while others will be downplayed or disappear. Which capacities will flourish will depend on the potentiality that the technology contains, but also on the context in which it is situated, and which may help to foster its further development or impose limits on it. Research contexts or user contexts may both contribute to the further development of capacities, but will never determine the form they are going to get: accident and chance will have an important role.

The comparison of a developing technology with a human life fits with the abandonment of the subject–object divide that became the common metaphysical background of scientific activities during the Enlightenment. Over these last decades many philosophers of technology have abandoned this metaphysical divide, such as Bruno Latour, Don Ihde and Peter-Paul Verbeek, who have all argued—in differing ways—how technologies do not figure as passive material objects, but offer active contributions to human actions, interactions, relations and experiences, thus making them a kind of moral subjects. Even though technologies do not have consciousness and do not reflect on alternative options for action, they resemble human moral agents in the sense that their role in interactions influences the value of human (social) life. Ascribing to them the role of an 'agent', however, presupposes that there is a kind of 'subject' or, in other words, 'something' to which such capacities can be ascribed. In the case of technologies that are being researched, however, this 'subject' is lacking. But the capacities are there to be investigated and evaluated. The 'something' about which we are talking as ethicists in the laboratory should not be talked about as a noun, but as a characteristic potentiality or activity, or as features. The predicates precede the noun to which they can be ascribed and invite or allow, and sometimes require, ethical reflection.

References

Collingridge, D. (1980) *The Social Control of Technology* (London: Pinter Publishers).

Daston, L. (2000) *Biographies of Scientific Objects* (Chicago: University of Chicago Press).

Ihde, D. (1990) *Technology and the Lifeworld* (Bloomington, IN: Indiana University Press).

Johnson, D. (2007) 'Ethics and technology 'in the making': an essay on the challenge of nanoethics', *Nanoethics*, 1: 21–30.

Latour, B. (1992) 'Where are the Missing Masses? The Sociology of a Few Mundane Artifacts', in Bijker, W. E. and Law, J. (eds) *Shaping Technology/Building Society* (Cambridge, MA: MIT Press).

Latour, B. (1993) *We Have Never Been Modern* (Cambridge, MA: Harvard University Press).

Russell, B. (1905) 'On Denoting', *Mind*, 14: 479–93.

Van der Burg, S. (2009) 'Imagining the Future of Photoacoustic Mammography', *Science and Engineering Ethics*, 15(1) 97–111.

Van der Burg, S. (2010) 'Ethical Imagination: Broadening Laboratory Deliberations', in Roeser, S. (ed.) *Emotions about Risky Technologies, Series International Library of Ethics, Law and Technology* (Dordrecht: Springer).

Verbeek, P.-P., (2005) *What Things Do: Philosophical Reflections on Technology, Agency and Design* (University Park, PA: Pennsylvania State University Press).

Verbeek, P.-P. (2011) *Moralizing Technology; Understanding and Designing the Morality of Things* (Chicago, London: University of Chicago Press).

6
Ethics on the Basis of Technological Choices

Xavier Guchet

Introduction

In the past few decades, the philosophy of technology has shifted from its 'classical' approach (focusing on how *technology* as a whole negatively affects *society* at large, and even the human condition) to a more empirical approach focusing on objects themselves (Brey 2010). Now philosophers are becoming more and more interested in the way concrete objects are designed in the laboratory in order to scrutinise how objects materialise norms and values, and introduce them into society. Philosophers of technology have shifted from their usual observation posture to a more active one; in other words, they have to more actively intervene in the design of technologies—they have to become fully-fledged actors of engineering processes. What about their traditional role as critics of current social organisation, and of its admitted set of norms and values? Are they in danger of losing their vigilance, which requires a certain 'distance'? Is the social role of philosophy now limited to 'accompanying' science and technology, and to providing an expertise for engineers and science policy-makers? This perspective is obviously unsatisfactory and raises the question of the role of philosophy when philosophers have to face such demands.

The empirical turn in the philosophy of technology means, above all, that values are not external to scientific facts. Philosophers are not the experts of values, while scientists remain the experts of facts: the need for a close collaboration between philosophers and scientists means that facts and values must be 'co-constructed'. But how does such an entanglement of facts and values occur? *Who* is legitimate in attaching values to facts? Is this attachment a separate process with regard to the research process itself or are they the same process? Does the value-shaping process come *after* the facts-building process—leading to evaluative approaches focusing on 'impacts', on the 'consequences' of techno-science on society, or are both processes deeply intertwined from the beginning? What exactly is the role of philosophy here?

Preliminary theoretical insights: Values and design

It is notable that ethical and social debates related to technology focus primarily on their 'impacts'. Most often, the *use* of technology and the prior *significance* the user gives to it come under scrutiny, but far less attention is usually paid to the *design* of the technological objects itself, before any kind of use is observable. Yet such a divide between design and values deserves to be challenged. Technological design is not neutral: it is value-driven. Ethical and social issues raised by technologies are not only related to their applications, to the way people behave with regard to them: ethical and social issues are framed upstream, in scientific and technological design choices (Verbeek 2007).

This approach to technology is not new. Perhaps the most well-known approach is value-sensitive design (VSD), which focuses on the integration of moral values into the conception, design and development of information technology (IT) (Manders-Huits 2010). Design choices made by engineers have moral and social consequences: VSD is a proactive method devoted to the analysis of these close relationships between technological design choices and ethical–social–political issues. However, VSD continues to focus on the *use* of technology—more precisely, on how a specific design is expected to affect current representations, expectations and social behaviours as soon as a technology is used.

An apparently more suitable approach, labelled constructive technology assessment (CTA), was developed in the Netherlands in the 1980s. CTA is part of the family of technology assessment (TA) approaches, but it is original because it focuses on the design process itself. While in traditional TA approaches, technology is taken as given, CTA focuses on the design of technologies itself. CTA militates against the so-called 'two-track regime' (Schot and Rip 1996), which opposes promotional activities related to technology on the one side, and control and regulation of technology on the other side. CTA aims at intertwining both; control and regulation processes must be closely intertwined in innovation processes. However, CTA continues to envision technological design through its potential 'impacts' on society.

Yet this unquestioned focus on 'impacts' must be challenged, for at least two reasons. First, technologies can be of major interest, from a philosophical point of view, even though their 'impacts' on nature and society cannot be anticipated. Let us remember what Bergson claimed 100 years ago: a technology, such as the steam engine, has deeply transformed European societies; it has changed our material conditions of life, our mentality, and even our perception of space and time. However, wrote Bergson, such effects started to get noticed long after the emergence of this technology. To be sure, acknowledging this existing gap between the social success of a technology and the awareness of its effects on society does not mean abdicating any serious attempt at evaluating present choices; however, evaluating them only with regard to their potential 'impacts' supposes that what is relevant for us today will still be relevant tomorrow. But how to make sure that the mental categories through which we experience our world will not become obsolete in the near future, such an obsolescence making the world of tomorrow, provided with present technologies, impossible to anticipate? The term 'impact' has a mechanistic significance, and in mechanics something must be unchanged (momentum, energy or everything else) for changes to occur. However, history does not rely on mechanistic principles. We cannot a priori make the divide between what technologies will leave unchanged and what they will 'impact'. So there is a need for another kind of entanglement of technology and society, alternative to the widely accepted focus on the 'impacts' of the former on the latter.

Second, ethical issues are often addressed in general terms, such as human dignity, social justice, inequality between north and south, privacy, environmental sustainability, etc. Relevant issues are not shaped with regard to specific technological design; a large amount of research is ethically and socially evaluated with few general criteria. That's why many official reports related to ethical and social issues of science and technology seem so monotonous. To overcome this situation, Paul Rabinow and Gaymon Bennett militate for a better entanglement of ethics and technology through more pragmatist modes of intervention by anthropologists and philosophers into design processes (Rabinow and Bennett 2007). They cite 'equipment' as being a better entanglement between values assessment and engineering practices. Rabinow and Bennett explain, however, that the criteria of such a close evaluation of design practices and technological objects relate to the concept of 'flourishing existence'. Does such and such technological object favour a 'flourishing existence' or does it not? The problem is then to define what a 'flourishing existence' is—a task that appears to be very challenging. Who is competent to define what a 'flourishing existence' is? Rabinow and Bennett's diagnosis appears relevant, but their concept of 'flourishing existence' cannot easily guide close collaborations between philosophers and engineers.

Yet we are not completely deprived of resources to take up this challenge. I claim that pragmatist philosopher John Dewey's theory of value is of major interest here (Dewey 1939). Dewey addresses a core issue: when existing values and associated goods are challenged, how are *new* values and *new* goods created? A theory of value is what Dewey calls a theory of 'valuing', of attaching price to something, i.e. a theory that explains and regulates the production of new goods and values in society.

Dewey's theory of value deserves attention in so far as it leads to a shift in the philosophy of technology: technological objects must be 'valued' not only as 'means' for reaching predefined ends, but also as things that take place in the world.

To be sure, there is a great difference between assessing the 'impacts' of a technology on society and 'valuing' the world it will contribute to reshaping. For instance, we can address the 'impacts' of the gasoline engine car on nature and society, we can also make the divide between 'positive' and 'negative' 'impacts', we can then aim at reinforcing the formers and reducing the latters, e.g. carbon dioxide emissions. However, we can also ask if we want to continue to live in a world in which individual cars organise our daily lives. It is not only a question of positive or negative 'impacts' of technologies; it is a question of 'valuing', of giving price to a certain world in which we must accept to live if we 'value' a certain technology. Assessing the 'impacts' of carbon dioxide emissions can lead to considerations about electric vehicles as a panacea, but in so doing we avoid questioning the 'value' of a technological solution that does not challenge the general organisation of contemporary societies. In this different perspective, gasoline engine cars are not only considered as 'means' for transportation, with positive and negative effects; they are also considered as things that have contributed dramatically to shaping the world we inhabit. Even if all negative 'impacts' of the gasoline car were avoided, the value of the world it co-creates would still be at stake.

So the empirical turn in the philosophy of technology means, above all, a shift from an impacts-centred approach to an objects-centred approach. The human world differs from animal worlds in so far as the former is shaped by objects. Strictly speaking, objects are not elementary parts of the human world: they are what French philosopher Maurice Merleau-Ponty called 'total parts' of it (*'parties totales'* in French) or 'dimensions' of it. They are parts of the world, but, at the same time, these parts can reshape the human world as a whole and give rise to a new experience of it. Objects shape the world in which they have a place. They give it its consistency. Undoubtedly, objects are parts of the world, but the world is also part of the objects. As Arie Rip emphasised (Rip 2009), technology paves the way for a 'prospective ontology', i.e. the objects of the present world (the 'furniture of the world' as Rip says, quoting Russell) contain prospective elements that embody expectations which open the way to further developments and direct the evolution of the world.

It is time for scientists to see their objects of research as 'prospectives' of a future valuable world. I present here a case study that illustrates an encounter between scientists and philosophers around a definite object. The purpose of this encounter was to 'value' the chosen object with regard to spontaneous and unquestioned mental categories through which we experience our world.

The case study: The remake of a bacterial flagellar motor in the laboratory

The case study belongs to the field of nanobiotechnology; it is about the artificial synthesis of the flagellar motor of bacteria. This project was launched by two researchers who belong to the Nano Bio Systems (NBS) group. The NBS group is part of an important French laboratory, the Laboratoire d'Analyse et d'Architecture des Systèmes (Centre national de la recherche scientifique), which was created in 1968, and is devoted to the analysis and building of systems. The NBS group consists of 8 permanent researchers, mostly physicists, and about 20 PhD students and postdoctoral fellows. More specifically, the group is interested in nanosystem applications for biology. It develops research in five areas: nanoscale slit biosensor, piezoelectric electromechanical nanosystems, molecular nanomotor, nanowires for high resolution thermometry and biodetection by means of electromechanical microsystems. The aim of the group is both to provide biology and medicine with new technologies, and to develop more fundamental research. The molecular nanomotor project is less applicative than other projects; above all, it is devoted to better understanding the structure and the functioning of the bacterial flagellar motor, even though some applications have been predicated by scientists for drug delivery, in particular.

The molecular nanomotor project was founded by the French Research Agency (Agence Nationale de la Recherche; ANR). The NBS group obtained a first project devoted to the nanomotor, FLANAMO

(2006–2009). At that stage, the nanomotor project involved a PhD student. A second project, FLANAMOVE (2010–2013), was founded jointly by the French research agency and the American National Science Foundation. The PhD student involved in the FLANAMO project obtained his PhD in June 2009, and left the NBS group for a postdoctoral fellowship in a research group devoted to biotechnology and synthetic biology at the University of Minnesota—both groups have become partners in order to improve the method for mass producing the proteins of the flagellar motor. The NBS team has recruited another PhD student (for the FLANAMOVE project), a biologist who is interested in working on this improved method. A second PhD student will work on the best way for these proteins to be adsorbed on the artificial surface.

A famous Modern credo says if we want to fully understand something, we have to (re)make it. The NBS group actually plans to artificially synthesise the tiny biological motor in order to get to know it better. This biological motor is of major interest for both scientific and technological reasons. On the one hand, it is one of the more complex existing protein machines. Understanding its functioning would be of great interest to biologists. On the other hand, this tiny motor has fascinating properties. It has no rival performance in human-made technology. Biologists have studied it for more than 30 years, but, up to now, they have not been able to propose a completely suitable model of its functioning. This situation has proven to be very challenging for engineers, as they cannot plan to build the nanomotor on the basis of a blueprint. The challenge of the NBS group is to overcome this problem, and to propose a technological protocol that would allow scientists to artificially synthesise the protein motor, even though no satisfactory model of it is available. This protocol relies on various key technologies: protein production by means of genetically-modified bacteria, surface chemistry, soft lithography, self-assembly and dynamic imaging (by means of an atomic force microscope). The proteins are mass produced and purified. Soft lithography is used to prepare an artificial surface mimicking the natural membrane of the bacteria. The proteins are printed on the prepared surface by means of a very special technique called 'molecular stamping', or soft lithography, which was first developed by chemist George Whiteside's group. The scientists involved in the nanomotor project hope that the proteins will be 'deluded' and will self-assemble on the artificial mimicking membrane the way they do in living bacteria. So they hope to see the prepared surface self-manufacture the tiny motor.

During the first phase of the project (i.e. the FLANAMO phase), the involved scientists faced two difficulties: first, the hope that the motor would self-assemble protein by protein on the artificial surface proved to be unrealistic; second, the mass production of certain proteins, especially transmembrane proteins, proved to be very challenging. In order to overcome these difficulties, the French group decided to take advantage of a method for engineering proteins that was developed at the University of Minnesota, where the PhD candidate from the first project (FLANAMO) was recruited as a postdoctoral fellow. This method is based on the design of a 'protocell'. The 'protocell' is an artificial lipidic vesicle—a bioreactor in which the cellular components involved in the protein synthesis are injected. The scientists expect to see the bricks of the nanomotor self-assembling in the protocell: these pre-assembled bricks will be extracted from the 'protocell' and printed on the artificial surface using the same 'molecular stamping' technique. In this new phase of the project, which consists of linking nanotechnology with synthetic biology, both previous difficulties are supposed to be overcome.

As the FLANAMO project took off, a group of philosophers launched a project (BIONANOETHICS), founded by the ANR, devoted to the analysis of philosophical issues related to nanotechnologies and biotechnologies. A PhD student, Sacha Loeve, and I (I was the postdoctoral fellow on the project) decided to focus on a particular class of nano-objects: molecular machines. This choice was motivated by two major considerations. First, the topic of the machine appeared to be strongly linked to the Grand Vision of nanotechnology—see, for instance, the great concern for Drexler's universal assemblers. We wanted to compare such visions and discourses with laboratory practices: Do molecular machines built and studied in laboratories really fit these visions and discourses? Second, we wanted primarily to tackle nanotechnology, neither as philosophers of science nor from a science and technology studies point of view, but as philosophers of technology. Yet the machine is of major importance in the philosophy of technology. We thought it would be relevant to bind nanotechnology to this conceptual history related to machines.

The encounter between us and the NBS group in 2006 was our initiative. We thought we would find an excellent example of the molecular machine here: the molecular nanomotor project. Over a few years, we met three times for interviews related to this project: in November 2006 (collective interview with some of the permanent researchers and some of the PhD students, but the postdoctoral fellow was not present); in October 2007 (individual interview with the PhD student involved in the project); in June 2010 (collective interview and individual interview with the professor responsible for the project). Prior to meeting the group for the first time in 2006, we sent them two philosophical texts in order to prepare the collective discussion. Both texts—an extract from Hannah Arendt's *The Human Condition* and a text written by Sacha Loeve—were related to the nature/artefact issue. Our research question was: How to qualify the remade bacterial flagellar motor if the project proves to be successful? Would it be natural or artificial? We wanted to give rise to a discussion on this topic with scientists.

The nature/artefact issue has proven to be very challenging because it appears to be often both over- and under-determined in evaluations related to new technologies—over-determined with respect to the constant reference to religion, to the idea that an intelligent design (ID) is at work in nature, i.e. to strong social and moral statements. Owing to its complexity, the bacterial flagellar motor has prompted vivid debates about ID. A clear-cut opposition is at stake here: either nature is the result of an ID and proves to be sacred, or nature is due to blind mechanisms, deprived of any intention and it is equivalent to artificial things. This simplistic alternative may prove useful, but, unfortunately, it does not fit scientists' own evaluations. The professor responsible for the molecular nanomotor project claims that each living being, including humans, is enclosed in a whole set of relationships. He summarises this thought in a very challenging statement: it is not necessary for a humanist to believe in God; however, someone who believes only in man is not necessarily a humanist...

Living beings are not the fruit of an ID; however, neither are they isolated things that we could manipulate: they are part of a whole and considering them to be purely artificial would be absurd. During our first meeting in 2006 the professor gave an example: a colleague of his presented a research programme at a workshop that consisted of genetically modifying a fly so that it would spontaneously manufacture a camera! For the professor we interviewed, the problem at stake here does not rely on the attempt at genetically modifying flies—flies are not sacred! However, it is scientifically absurd to claim that a fly could be redesigned genetically to produce a camera or a microprocessor. The difference between what is manufactured and what is generated biologically cannot be blurred completely. So a statement on the nature/artefact issue, or on the biological/technological divide, is not purely speculative: it has practical consequences, it affects the direction of research programmes and it can be a strong argument emphasised by scientists in their evaluations of colleagues. However, this nature/artefact issue, while it is over-determined by moral and social debates, is also under-determined: interviewed scientists of the NBS group confessed they had no adequate concepts allowing them to overcome the strong alternative between sacralisation of nature and radical artificialisation of nature.

At this point the philosopher can help. As philosophers of technology, we do not consider our role as being to provide scientists with a set of moral principles, drawing the line between the permitted and the forbidden. Our role is to provide scientists with concepts that could allow them to better qualify the objects they deal with: if such objects are neither purely natural nor purely artificial how can we talk about them? And what are the relevant issues at stake in manipulating them? On which conceptual basis could we claim that some manipulations of living beings are biologically, and therefore scientifically, absurd?

Further discussions between philosophers and researchers of the NBS group were related to this challenging nature/artefact issue. The BIONANOETHICS group organised a two-day international workshop in January 2008. Philosophers wrote papers that were mostly related to this issue, and the papers were commented on by scientists.

So our approach to the interaction between scientists and philosophers relies on the shared reading of texts—of philosophical texts, but also of scientific texts. Indeed, from the outset, in order to highlight the genuine values attached to their own project on molecular nanomotor design, French scientists compared it to rival research projects. This is another important characteristic of our approach: the combination of scientific and technological choices, on the one hand, and values, on the other hand, become clear as soon as a comparison is made with alternative projects; the values attached to a defined research are differential. A technological design becomes valuable in regard to alternative design strategies.

In addition to the 'fly-camera' project, scientists mentioned two rival research projects. The first of them was conducted by a Japanese team. This project aimed to harness bacteria in order to move a micromotor (this paper was presented by scientists and discussed with us during one of our meetings; Hiratsuka and alii 2006). The starting point of both teams was the same: both began from the observation that it has proven impossible to synthesise biological motors from scratch, protein by protein. Just as in the previous project, the Japanese scientists were unable to explain the mechanism of the bacterial motion in detail. They were also unable to rebuild the bacterial motor from its components.

It became clear that harnessing natural mechanisms was a condition for success. The challenge of the Japanese team was to use the bacteria themselves in order to give power to a useful hybrid micromachine. The micro-rotary motor is composed of a tiny rotor, a sort of toothed wheel which is the mobile part of the motor and is made of silicon dioxide; the wheel is connected to a silicon track on which gliding bacteria called *Mycoplasma mobile* circulate. The role of the bacteria is to 'push' the wheel to make it turn so the bacteria activate the motor. Bacteria are 'fed' glucose, which turns out to be the fuel of this tiny motor. The major technological challenge consisted of modifying the surface of the bacteria, and in preparing the artificial track so that bacteria would attach to the surface of the track and would move in a given direction, forcing the rotor to turn.

The second project was Craig Venter's attempt at synthesising the first artificial self-replicating bacterial cell, provided with a completely synthetic genome. Venter's team succeeded and published the results in *Science* (Venter and alii 2010). Venter's team designed, synthesised and assembled an entirely artificial genome of 1.08 mega-base pairs, then transplanted it into bacteria that had been previously enucleated. So, Venter's team created the first synthetic bacteria to be controlled by a synthetic chromosome. The new bacteria, comprising the entirely synthetic and digitally designed genome, have proved self-replicating. This news has been considered of major importance in so far as it appears to be a crucial step toward artificial life.

French scientists evaluated the alternative projects with regard to both technological skills and to moral expectations. The Japanese research is considered interesting, even though the results are not completely convincing: the tiny motor powered by bacteria turns at a very slow rate and has not proved efficient with regard to existing microelectromechanical systems. However, the idea was appealing and the design strategy was inventive. French scientists, however, point out a moral issue: the Japanese project involves 'enslaving' living beings—they use this term—which is of moral concern, even though such beings are at the bottom of the evolutionary scale. Scientists have nicknamed the Japanese micro-rotary motor the 'bacterial mill'—as if bacteria were 'slaves' powering a mill. Such evaluation is surprising in so far as French scientists harness bacteria too: they genetically modify bacteria to provide them with a large amount of proteins. Actually, they also use the term 'enslave' (in French, 'mettre en esclavage') when they talk about their genetically-modified bacteria. So what's the difference? Dialogue between scientists and philosophers can help to clarify this issue.

In both cases, living beings are harnessed; however, in the French case, bacteria are considered reservoirs of raw materials: they provide scientists with proteins, just like animals traditionally provide hunters with the proteins they need. In the Japanese case, bacteria are not considered reservoirs of raw materials, but 'slaves' in the sense that they are harnessed to achieve human tasks, just like oxen, for instance. So what is the problem? We harness oxen; why not bacteria? A possible interpretation is of cultural concern: before harnessing oxen or horses we had to domesticate and tame them. We had to consider them our partners, full members of our human world. The Japanese design strategy consists of 'enslaving' living beings without domesticating and taming them. Such a strategy may be of important significance with respect to the traditional inherited cultural framework that, for thousands of years, has shaped our relationships with animals—a framework that appears challenged here. This short circuit—of avoiding the domestication and taming of animals before harnessing them for useful tasks—proves to be of possible moral concern. A sort of chiasm seems to be at stake here. Familiar domesticated animals, such as bovine, ovine or poultry, have become a great matter of ethical concern, especially in so far as they are industrial 'materials'. Such ethical issues are related mostly to the animals' well-being. While we do not spontaneously consider the harnessing (the 'enslaving') of such animals in order to achieve human tasks, for instance for farming, of moral concern—provided that their well-being is respected, i.e. that they are considered members of the human world—we morally condemn industrial breeding in which animals are considered pure reservoirs of raw materials and are mistreated. On the contrary, we do not morally condemn the use of bacteria as reservoirs of raw materials (e.g. of proteins); however, there is something strange in trying to harness them as if they were oxen—this strangeness can be related to the fact that bacteria are not considered partners of human beings, full members of the human world, even though they are ubiquitous.

So the comparison between both projects—the Japanese one and the NBS one—challenges our relationships with animals; it also challenges the spontaneous hierarchy we make between animals that are considered partners and members of our human world, and animals that are not. Furthermore, as we go down the taxonomic hierarchy (humans being at the top of it), we find animals that seem to be deprived of any moral concern: insects or bacteria. Is such a convergence between a taxonomic hierarchy and a moral hierarchy legitimate? Is there any valid reason to refuse bacteria that which we grant mammals—the commitment to respect their well-being according to their own needs? No doubt that such a chiasm involving domestic animals on the one hand and bacteria on the other paves the way for questioning our moral concerns related to living beings that surround us.

Craig Venter's project is controversial. French scientists blame Venter of having overestimated the importance of his results. Undoubtedly, these results are very impressive from a technological point of view: synthesising an artificial genome of more than one million base pairs was an incredible challenge. However, Venter's success cannot be viewed as the decisive step towards creating artificial life: he did not artificially synthesise a living cell; he took an existing cell (*Mycoplasma capricolum*), he enucleated it, and he introduced the artificially designed and synthesised sequence of nucleotides into it. The redesigned cell proved self-replicating, which is an important result—perhaps a major step towards the knowledge of what a 'minimal genome' could be; however, this does not mean that life was artificially created. While French scientists want to *both* achieve a fruitful technological synthesis protocol and to better understand the functioning of the bacterial flagellar motor, Craig Venter's project is considered of poor scientific interest: it is only tinkering, 'bricolage' in French—undoubtedly a clever 'bricolage', but nothing more than a 'bricolage'. Furthermore, this 'bricolage' may prove dangerous. Indeed, French scientists say, that nature has provided living beings with important abilities to adapt to environmental modifications for billions of years. What about a living being—Venter's synthetic bacteria—which has been designed in silico and which is therefore deprived of any relationship with an evolutionary process? Either such an artificial living being self-replicates so that it continues to do what its designers wanted it to, but, in that case, it will quickly prove incapable of any adaptive behaviour and will perish, or it acquires genuine adaptive abilities and will, as a consequence, escape its designers' programme. By contrast, the NBS group underlines the scientific challenge of its project (it is not only 'bricolage') and the modesty of it, which is of moral significance: they do not attempt to make life from scratch in the laboratory.

The NBS group project is, however, of philosophical concern. Such concern short circuits currently addressed issues, such as the nature/artefact dichotomy, the issue of the instrumental approach to living beings and the moral questions attached to the attempt to artificially create life. In this case study, such general topics seem to be misleading. The comparison between the NBS group project, the Japanese one and Craig Venter's claim has undermined the relevance of the second and the third issues. The first one—the nature/artefact problem—is also of poor interest here; undoubtedly, the molecular nanomotor project does not aim at 'artificialising' nature in so far as it accurately consists of leaving nature to operate on its own. Rather, scientists claim that if they succeed, the synthesised molecular nanomotor will prove identical to the biological one; so instead of 'artificialising' nature, the project opens the way for 'naturalising' artefacts. Moreover, this statement is not completely satisfactory: there is no evidence that the synthesised motor would be exactly identical to the bacterial flagellar motor. As it is strongly related to the alternative 'artificialisation' of nature versus 'naturalisation' of artefacts, the nature/artefact issue does not seem relevant in this case study. We need another approach.

At the edge of the nature/artefact issue the novelty of the molecular nanomotor project lies in the blurring of a clear-cut divide that ancient Greeks delineated and that have proved very meaningful in Western cultures—a divide between two kinds of technological activities. In his *Physics*, book 2, Aristotle distinguished between the manufacturer and the pilot or the farmer: both have technological skills, but these skills are very different. While in the first such skills consist of transforming matter to make an object, in the second they consist of making an arrangement with natural processes in order to achieve human goals (for instance, carrying goods).

Yet, if the NBS group achieve the molecular nanomotor project, the frontier between objects that are man-made (such as a motor) and objects that are 'piloted' or harvested (such as a plant) will be blurred. Roboticist Rodney Brooks claimed a few years ago that very soon, instead of planting a tree, waiting for it to grow, cutting it down and manufacturing a table, we would be able to make the table grow. The molecular nanomotor project fits this statement; it paves the way for the complete victory of harvesting over manufacturing.

So what? To what extent would such a blurring of manufacturing and 'piloting' natural processes be of moral concern? Can we imagine a world in which tables and motors would not be manufactured any more, but would grow spontaneously? Let us make a few comments.

Until now we have built factories to manufacture everything we need, from tables to motors. In the last few decades, industrial development has proved to be challenging for nature and its equilibrium. From now on, however, in the new perspective introduced by the blurring of manufacturing/'piloting', nature as a whole would be considered a giant factory. Instead of reducing our ecological footprint on nature, we would extend the factory model to the whole of nature. No doubt such a perspective would be of great concern to the field of environmental ethics: What would the ecological consequence of Rodney Brooks' claim be?

Furthermore, the evolution of humans—palaeoanthropologists have proved—has been closely related to the transformation of matter by a tool-enhanced hand. So what could be the anthropological significance of the possible divorce between man and manufacture? What about the situation of humans in the world if they give up transforming matter by means of tools and machines, i.e. what about the future of humans if they give up one of the main activities by which they became human? The question at stake here is not only related to potential ethical and social 'impacts' of the laboratory molecular nanomotor if the project succeeds; it is related to the challenging of fundamental categories through which the world has been made human. French palaeoanthropologist André Leroi-Gourhan (1911–1986) called 'regression of hand' (in French, *régression de la main*) the likely consequence of such a divorce between humans and manufacture, i.e. the loss of direct contact with matter (Leroi-Gourhan 1993). The 'regression of hand' would challenge the biomechanical basis of the human body, such as upright posture and teething. The contemporary literature related to post-humanism mostly emphasises the idea of 'enhancement'—enhancement of physical and social performances of the brain, etc. However, according to Leroi-Gourhan, the redevelopment of biomechanical features that have made us human is also at stake here. Obviously, the consequences of such an evolution cannot be anticipated, but Pixar Animation Studios produced a wonderful film entitled *Wall-E* a few years ago, the story of a little robot programmed to clean a devastated earth full of rubbish. Humankind has left this uninhabitable earth for a spaceship that provides humans with everything they need—as if commodities grew of their own. However, in this new form of life, human beings have lost their upright posture and their teeth. So this beautiful story for children fits palaeoanthropology lessons.

I do not claim that the molecular nanomotor project, if it succeeds, would make us all lose our teeth within a few years, or even a few decades! However, the role of philosophers is to propose alternative timescales at which phenomena make sense: at the time scale of evolution, as Leroi-Gourhan claims, the regression of hand would be of great importance to humanity.

Last comment: a dramatic divorce between humans and manufacture would dramatically drive users apart from any possible knowledge of what they use. The lack of common knowledge about technology today has become a leitmotiv. The manufacturing/piloting blurring would most likely widen this gap. While we can still associate a table, or a motor, with the set of operations that produce it, a 'grown' table or motor would suggest that something like magic governs our societies. Obviously, there would be a lot of knowledge and technological skills in the 'grown' table or motor—no doubt the molecular nanomotor project will prove to be of a high scientific and technological level! However, this raises the question of whether the gap between scientists in their laboratories, on the one hand, and lay-people deprived of any knowledge about the production of their commodities, on the other hand, would not be dramatically broadened. As French philosopher Henri Bergson claimed, while science loses ground, magic still lies in wait...

These comments do not aim to delineate the line between the permitted and the forbidden. Philosophers are not censors. Their role is not to anticipate 'impacts' of science and technology—I agree with Bergson that such anticipation mostly proves impossible. How could we anticipate the 'impacts' of the molecular nanomotor project at the palaeoanthropological timescale? However, even though anticipation is often deprived of any sense, it is possible to point out potential relevant meanings of research projects and to discuss them with scientists. The collaboration with the NBS group is on-going; the comments made in this chapter deserve further discussion between philosophers and scientists.

Conclusion

In the context of the empirical turn in the philosophy of technology, philosophers seem to compete with each other in getting as near as possible to scientific and technological practices (Nascimento and Polvora 2011). Undoubtedly, such a shift has proved fruitful: philosophy of technology appears less distant and more concerned with actual and relevant issues related to concrete research. However, this recent tendency deserves accurate reflection. Are philosophers becoming new kinds of 'experts', working for policy-makers who fund them generously through research projects? The keen interest in philosophy today, and especially in ethics, has contributed dramatically to treating philosophers as mediators between scientists/engineers and society. As they are considered 'experts' in values, philosophers are now committed to feeding scientific policy with their expertise. So getting closer to technological practices is two-edged: it paves the way for better addressing relevant issues related to science and technology, but it also encourages philosophers to join a strategy in making the 'social acceptance' of technology easier. How to get close to technological practices and objects without ceasing to be external to such strategies? In this chapter I have claimed that a possible solution consists of shifting the focus away from assessing the 'impacts' of technology on society. Philosophers can bring something more than an 'expertise' or a set of principles in order to increase 'positive impacts' of technology, and to reduce 'negative' ones; rather, they can provide scientists with a specifically philosophical *ethos*, consisting of helping them make relevant divides where frontiers are blurred and in blurring too clear-cut admitted divides. Getting closer to scientists and engineers in their laboratories should mean, above all, getting more and more distant with regard to the current trend of recruiting every social force—including imagination and desire—for the scientific and technological development to continue.

References

Brey, P. (2010) 'Philosophy of Technology after the Empirical Turn', *Techné: Research in Philosophy of Technology*, 14: 1.

Dewey, J. (1939) *Theory of Valuation* (Chicago: University of Chicago Press).

Hiratsuka, Y. and alii (2006) 'A Microrotary Motor Powered by Bacteria', *PNAS*, 103(37): 13618–23.

Leroi-Gourhan, A. (1993) *Gesture and Speech* (Cambridge, MA: MIT Press).

Manders-Huits, N. (2011) 'What Values in Design? The Challenge of Incorporation Moral Values into Design', *Science and Engineering Ethics*, 17(2): 271–87.

Nascimento, S. and Polvora, A. (2011) 'Towards the Participation of Social Sciences and Humanities in the Practical Realms of Technology', *The International Journal of Technology, Knowledge and Society*, 7(2): 61–72.

Rabinow, P., and Bennett, G. (2007) *From Bio-Ethics to Human Practices: Assembling Contemporary Equipment* (Cambridge, MA: Daedalus).

Rip, A. (2009) 'Technology as Prospective Ontology', *Synthese*, 168: 405–22.

Schot, J. and Rip, A. (1996) 'The Past and Future of Constructive Technology Assessment', *Technological Forecasting and Social Change*, 54: 251–68.

Venter, C. and alii (2010) 'Creation of a Bacterial Cell Controlled by a Chemically Synthesized Genome', *Science*, 329(5987): 52–6.

Verbeek, P. (2007) 'Design Ethics and the Morality of Technological Artifacts', in Vermaas, P. E. and alii (eds) *Philosophy and Design: From Engineering to Architecture* (Dordrecht: Springer).

7
Environmental Ethics in an Ecotoxicology Laboratory

Fern Wickson

Introduction

The move towards engaging ethicists on the laboratory floor as an approach to encouraging responsible research and innovation has focused largely on engaging with scientists actively involved in technology development. The case study described in this chapter adopts a slightly different orientation. The laboratory at the centre of this case study is not engaged in research oriented towards creating technological development, but rather in conducting research on the potential risks posed by technological development. More specifically, it is a laboratory conducting ecotoxicological research to understand the environmental effects of genetically-modified organisms (GMOs) and their associated pesticide regimes. As such, it performs research specifically intended to inform decision-making on new and emerging technologies, and support responsible use and applications of such technologies. This means that, although it is not research directly aiming to develop new technologies, it is research that actively shapes those technologies through the way it informs industry developments, risk assessment, regulation, governance and public opinion.

Environmental toxicology, or ecotoxicology,[1] began as a field of research to assess the effect that different chemicals (and particularly those from industrial waste discharges) would have on the environment (Cairns Jr 1995). The primary method to perform such research is to expose a set of standardised indicator species to selected individual chemicals in varying concentrations over short periods of time, documenting effects on factors such as mortality, growth, reproductivity or, less commonly, behaviour. The results from such experiments are then extrapolated using safety factor multiplications to infer the level of potential harm posed by the chemical in question for all non-target organisms. This type of research forms the knowledge base for processes of environmental risk assessment, which, in turn, represents the main decision-aiding tool for regulatory authorities.

The toxicological approach to understanding environmental harm is now extended to the environmental effects of new technologies, including those related to GMOs. The use of toxicological approaches for understanding the environmental effect of GMOs has, however, been contested owing to concerns relating to the adequacy of transferring a paradigm developed to understand environmental effects from chemicals to those from living organisms (Andow and Hilbeck 2004). This is not only because of the dynamic and reproductive possibilities of living organisms, but also owing to questions surrounding the relevance of a universal set of indicator species for understanding environmental harm, particularly for specific local contexts and conditions. The indicator species that have been standardised for ecotoxicology and environmental risk assessment have been chosen because of their ease of cultivation in a laboratory, their chemical sensitivity, their genetic uniformity and their wide availability, rather than, for example, their ecological functionality, their social significance or their relevance to the geographical location in question (Andow and Hilbeck 2004; Chapman 2002). That said, the knowledge developed through ecotoxicological models and methods remains widely accepted, and, indeed, required for regulatory decision-making on new and emerging technologies.[2]

Toxicology generally, and ecotoxicology more specifically, represent a particular form of science that is oriented specifically towards informing policy, which Jasanoff (1990) refers to as a 'regulatory science' and Demortain (2011) calls an 'evaluative science'. These are sciences conducted specifically for the purpose of meeting legally mandated regulatory standards. While ecotoxicology plays a formal role in political decision-making on emerging technologies through providing the necessary knowledge base for established decision-making processes, such as environmental risk assessment, it is also important for governance more broadly. This is because ecotoxicology is both developed and used by other actors, such as industry and environmental non-governmental organisations, as evidence for assessing the desirability of pursuing particular technological trajectories. Ecotoxicology therefore plays a significant role in shaping both social and political understandings of environmental harm relating to new and emerging technologies, and influences how these technologies are understood, managed, controlled and governed within the industries that are developing them, the policy circles that are regulating them and the societies that are using them. In this way, it can be seen as a science oriented directly towards facilitating safe and responsible innovation, and therefore an interesting one to engage with and analyse in any process of ethical technology assessment.

The other, somewhat unique, focus of the case study described in this chapter is that although I was open to and encountered a range of ethical issues related to science, policy and responsibility in the ecotoxicology laboratory, there was a specific interest in investigating how questions of *environmental* ethics were manifest and approached by the scientists involved. Therefore, although other relevant ethical issues encountered on the laboratory floor are briefly pointed to at the end of this chapter, the focus of the discussion here is on the major fault lines of debate within environmental ethics and how these particular issues are conceptualised, encountered and handled in the research processes of an ecotoxicology laboratory.

Context: The laboratory floor

The research in this chapter is part of a project funded by the Norwegian Research Council to examine how environmental ethics and ecotoxicology may be integrated for responsible ecological governance of emerging technologies. The project was funded within a call specifically seeking 'integrated projects' in which social scientists would engage directly with processes of technology development and be embedded in laboratory research. In this way, the call was very much aligned with the philosophy and ideas of ethical technological assessment outlined in this book.

The project for which I applied, and from which the work in this chapter stems, actually has nanoparticles for environmental remediation as its case study, with partners in the project including both scientists developing these particles and those performing ecotoxicological research on them. While the first year of the project was planned as time for familiarisation with environmental ethics literature and debates (with laboratory engagement on the case study to be conducted in later years), my host institute (GenØk Centre for Biosafety) offered the possibility to begin engaging in an ecotoxicology laboratory immediately. Although the ecotoxicology laboratory at GenØk was focused on biotechnology rather than nanotechnology, the opportunity to familiarise myself with the practice of ecotoxicology and engage in laboratory work immediately (despite its focus on a different case study) was too appealing to ignore. Therefore, what is described in this chapter represents only the first year of a planned three-year project engaging with ethics on the laboratory floor, and therefore a type of preliminary scoping exercise that will be further developed and expanded into other laboratories and other technologies in the coming years.

It is worth specifically noting at this point the significance of the fact that the engagement described in this chapter represented a collaboration between social and natural science departments of the same institute, and an institute with a long-standing commitment to the importance and value of interdisciplinary, holistic, research. The engagement in this case therefore took place between institutional colleagues with commitment on both sides to the importance and stimulating nature of such an interaction. In addition to this, being colleagues in the same institute also allowed for a lot of informal face-to-face time (e.g. shared lunches, coffee breaks, social events, etc.), which meant that all the collaborators in the initiative had time to get to know one another in a deeper and more personal way than may otherwise have been the case. This not only facilitated the building of relationships of trust and respect, it also allowed for the social and natural scientists to understand one another's perspectives in a broader context of personal history, social values and life activities. These types of relational factors and shared understandings may be more difficult to establish if ethicists engaging in laboratories are unable to share such informal and social spaces and, therefore, they arguably represent an important dimension of the nature of this particular collaboration.

The ecotoxicology laboratory I engaged in performs experiments using the model organism *Daphnia magna*. *Daphnia magna* (commonly called water fleas) are small, freshwater crustaceans, approximately 5 mm in length, which feed on bacteria, algae or other particles of plant biomass, and are a common food source for fish and other organisms in aquatic ecosystems. Daphnids typically reproduce parthenogenically, meaning that populations usually consist of females and their genetically identical or clonal offspring. However, under periods of environmental stress or change (e.g. limited food, altered temperatures) the females can produce male offspring and reproduce sexually to create ephippia or resting eggs. Resting eggs can tolerate adverse conditions and be stimulated to hatch when conditions are favourable again. Daphnids are model organisms for aquatic toxicology and are favoured owing to their genetic uniformity, ease of cultivation, their short lifespan and their role in aquatic food webs.

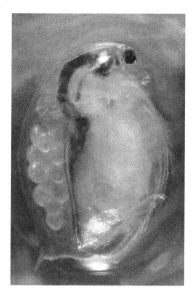

Figure 7.1   Daphnia magna.

During my time in the *Daphnia* laboratory at GenØk to date, I have been involved directly in two long-term, chronic exposure experiments. The first studied effects on *D. magna* when fed genetically-modified 'roundup ready' soya beans in comparison with conventionally and organically produced soya beans. The plant material used in this experiment was also tested for the different pesticides (and their concentrations) present in the soya beans, and therefore the effect of these chemical residues was also a component of the experiment. The second experiment studied the effects on *D. magna* when fed maize plants modified genetically to express an insecticidal 'Bt' toxin, in comparison with its non-genetically-modified 'isogenic' counterpart. My participation in this experiment extended over a period that also involved a range of small pilot experiments to investigate what would be the most interesting and relevant test materials and methods of exposure for the longer-term investigation. The research team involved in these experiments included the laboratory leader (hereafter referred to as LL), a PhD student (PhD), a Masters student on exchange from Brazil (M) and a short-term laboratory assistant (who participated in the soya experiment only).

My engagement in the *Daphnia* laboratory took place over a period of approximately eight months. When experiments were operating, I was spending 1–2 days a week working in the laboratory. Each of the two long-term experiments were conducted for 42 days, while the pilot experiments typically lasted around 21 days. During the period of my involvement, I also attended two major laboratory meetings: one during an experimental planning phase and another where the results from the maize experiment were being discussed. I also helped the laboratory staff plan and conduct a practical class using the *D. magna* model during GenØk's annual capacity-building course. While I conducted one formal in-depth qualitative interview with each of the research team members at the end of the engagement period (excluding the short-term laboratory assistant), during my participation in the laboratory work I also actively questioned the scientists and explored issues of relevance with them as they emerged, recording written reflections on such issues at the end of each day. The engagement ended when the laboratory had to be shut down for relocation; however, it is expected that it will be resumed when the laboratory is re-established in its new location.

In a previous research project where I was employed as a postdoctoral fellow, I experienced an unsuccessful attempt to conduct a laboratory ethnography. Aiming to avoid similar problems this time around, I had a specific intention to offer practical assistance to the scientists involved. That is, I wanted to create a kind of 'trading zone' (Galison 1997) in which the scientists involved would not just be objects of my study or actors that I hoped to influence, but rather would, in some way, directly benefit (in concrete terms and in their own endeavours) from my presence in the laboratory. This meant that at the risk of 'going native' I sought an active, hands-on role in the laboratory work. My capacity to participate in the laboratory evolved over the eight-month period as my skills and knowledge developed, and as the needs of the research team changed. In the beginning, for example, when all team members were involved in the experiment, I was engaged primarily in simple tasks, such as passing numbered beakers to the researchers and washing them as they were replaced in the experiment. I progressed to preparing glasses and feeding the experimental animals with the test material as I became more familiar with the laboratory, learnt skills in pipetting and gained the trust of the scientists involved. Finally, in the maize experiment, where it was often just myself and M performing the experimental routines, I took on responsibility for collecting data on reproductive endpoints, such as counting the number of juveniles produced by the experimental animals under a microscope and entering this information into a database.

Encountering environmental ethics in an ecotoxicology laboratory

In what follows, I aim to highlight how issues that are key fault lines of debate within environmental ethics were encountered, considered and negotiated in the ecotoxicology laboratory I engaged with. To help introduce and frame this discussion, I use the tool of an anecdote—a short story from the laboratory.

The Albert anecdote

This is an anecdote about an aquarium fish named Albert.

> Albert was a purple cichlid, with small aqua spots and stout reddish fins. He belonged to a scientist named Justin, who was a tall ecotoxicologist, with neither spots nor fins. Justin performed experiments on water fleas (*D. magna*)—tiny translucent crustaceans, with long, feather-fingered arms, and an ability to reproduce both sexually and asexually.
>
> Justin used daphnids in his experiments to try and understand the effects of agrochemicals and transgenic crops on non-target organisms. To produce the water fleas for his experiments, Justin had what he called his 'mother factories' (quickly rechristened 'mother populations' when questioned by what may have been perceived as an ethicist with feminist tendencies). Hundreds of daphnids were being produced, with each daphnid asexually generating ever more copies of itself. Some were needed for the experiments and some were needed to continue the mother factories, sorry, populations, but certainly not all of them. More were also being produced by the experimental animals, but became unnecessary once the reproductivity count had been documented. All of these animals needed to be destroyed somehow. But how?
>
> In the beginning, Justin was flushing—simply sending the little daphnids into the drains. It wasn't entirely comfortable though. It felt wrong somehow. Or, at least, it felt as though it could have been better. The daphnids were dying for a cause, a greater good, being sacrificed on the altar of knowledge. Therefore, when their job was done, when they had made their contribution to this quest, it

was thought that they should at least be given an ethical death. But what would that be? What would be an ethical death for a water flea?

Enter Albert. To be eaten by a fish was considered a 'more ethical' death. It just seemed more 'natural' somehow. But what is this 'natural'? What did it mean for Justin? And what would it come to mean for Albert?

Every day Albert would wait at the corner of his tank, facing Justin at his workbench. When he saw Justin come towards him, he would swim excitedly back and forth, waiting for his meal to be delivered. Albert was a rapid and effective hunter, and he ate well. Then, one day, the experiments were over. Justin had to write up his results and the laboratory had to move. As it would take a while to establish the new laboratory and write a PhD, all the instruments had to be packed away, the mother factories shut down, and somewhere found for Albert to live. Justin decided to take Albert home. After some weeks though, he noticed that Albert was not looking his best. His colours faded, his energy diminished, and then, one morning, Albert was dead. What was Justin to do? How should he dispose of our hero, the fish who had provided an ethical death for so many of his test subjects?

He could bury him, but that is what he did with the remains of all the fish he caught and ate from the sea. That didn't seem right. So what did Justin do with Albert? Why he flushed him of course! It just seemed 'natural' somehow, for a dead aquarium fish.

An anecdote is characterised by the way it communicates a reality or a truth that extends beyond the tale itself, and which, in some way, is broadly recognisable. My intention, therefore, is not to analyse this particular anecdote or it's central concept of 'naturalness' directly, as I think it stimulates reflection and speaks for itself that way. What I will do, however, is pick up some side themes from the story and discuss how they represent ways in which the ecotoxicologists encounter and need to negotiate two of the key fault lines of debate in environmental ethics: the issue of instrumental versus intrinsic value and how far to extend the boundary of our moral community.

Intrinsic versus instrumental value in nature:
Death of daphnids for the greater good?

One of the longest standing debates within environmental ethics concerns the question of whether the value of nature is purely instrumental or whether biological entities have value beyond their usefulness for

human endeavours and purposes. This debate concerns the motivation behind any quest for environmental protection and whether it
is to allow for on-going human use and enjoyment, or because the
organisms and systems involved have a value independent of human
aims and desires. While meta-ethical debates concerning the anthropogenic nature of value ascription are often entangled in debates over
intrinsic and instrumental value (e.g. see Norton 1991), the core question of interest is really over whether all value in nature is necessarily
anthropocentric, even if it is inevitably anthropogenic. That is, although
human consciousness may be considered the *source* of all value, this
does not necessarily mean that humanity is the only *locus* for intrinsic
value (Callicott 1986; Lee 2003). Within the literature on environmental ethics, the fault line of debate between instrumental and intrinsic
value in nature has been nuanced through arguments for strong and
weak forms of anthropocentrism (Norton 2003), objectivist and sub-
jectivist forms of intrinsic value (Hargrove 2003), various conceptions
of intrinsic value (Noer Lie and Wickson 2011; O'Neill 2003), distinc-
tions between intrinsic value and inherent worth (Taylor 1986) and
various arguments for the characteristics required for something to be
considered an end in itself (see Rolston 2003; Taylor 2003).

The scientists in the ecotoxicology laboratory I engaged with all
expressed, in some way or another, a belief in the intrinsic value of
nature, for example through statements such as 'nature has a right on
its own' (PhD). More directly though, they all lamented or expressed
a particular discomfort with the way nature is being 'exploited', 'pol-
luted', 'eliminated' or 'degraded' for distinctly human purposes. For
example: 'this utilitarian, resource-focused, anthropocentric way that
we have fucked everything up, exploited everything... I have known
since I was 3 years old that this was deplorable' (PhD). This lament over
anthropocentric forms of environmental degradation is in a clear way
what motivates their involvement in ecotoxicology research—a desire to
protect natural entities from further harm from human activity. Interestingly, this is arguably also one of the main aspects motivating the
development of ethical technology assessment and the quest for responsible innovation. This means that, in this case, both the scientists, and
I, as the engaged ethicist, shared a common desire to minimise environmental harm from new technologies. While none of the scientists were
originally trained in ecotoxicology (they have backgrounds in agron-
omy, marine biology and the ecology of invasive species), an interest in
environmental conservation and a desire to make a positive contribution was common across all of their activities. In addition, common to
each of them now is a specific interest in agriculture, with a scepticism

or ethical concern over industrialised models of food production, for example 'I am fairly anti-industry, I think, because I think they are more concerned with their pockets than with environment and people' (M).

Therefore, what the scientists in my ecotoxicology laboratory had to do was to negotiate their belief in a type of intrinsic value in nature, with their desire to contribute to conservationist causes through ecotoxicological research that necessitated the instrumental use of organisms. While not an issue that they reflected deeply over in their daily practice, for example 'Typically I brush this away, it is pushed down under the surface of everyday life, but you ask and then I think about it' (LL), when asked to articulate how they justify or think about their instrumental relationship with the daphnids, two expressed it as an ethical conflict that they typically repress rather than face and experience a feeling of discomfort over when questioned, while the other approached it through a virtue ethics perspective, emphasising a justification based on the instrumental use being performed with virtues of care and respect for the organisms involved.

> These animals have lives with complicated behaviours and senses, and we make them live in different concentrations of pesticides and potential toxins. They are used and exploited, and later killed as part of the knowledge system, which is for us... I could justify it by the importance of the information, but I don't like that justification. I am uncomfortable that they are treated as means, that they are used. It is hard to measure this against knowledge for healthier living. (LL)

> They were a tool, but I had a sense of care for them. I didn't want them to suffer... We were using them as tools, and manipulating them and putting them through controlled conditions, giving them only maize poor diets... it was a thought sometimes, but it didn't bother me, I thought I could see the good side of it, the research, which was the potential to provide insight and to elucidate some questions in relation to biosafety, and this is important to society, and I had some sense of care for them. I liked them. They're interesting beings... I would be gentle when I was manipulating them. I wouldn't treat them as if they just are things and they don't feel. I would have this in my mind. That they are beings and that they deserve some respect, especially because I am using them as I am using them. It is for a good cause I think, but that is not always justifiable. (M)

The question of demonstrating care and respect touches on another fault line of debate within environmental ethics, concerning how far the boundaries of our moral community should extend. The use of animals in research is a highly controversial topic and the question of animal welfare and rights is central within environmental ethics. However, the matter of how we define an animal, and just what kinds of animals are worthy of our respect and care is far from settled. This issue was also one confronted by the scientists on the laboratory floor.

The boundaries of our moral community: Do water fleas deserve an ethical death?

Within the legal frameworks for the welfare of animals used for research, there is no requirement that invertebrates such as daphnids receive any particular form of care. This is also the case in the Norwegian animal welfare act (Norwegian Government 2010), which uniquely states that 'Animals have an intrinsic value which is irrespective of the usable value they may have for man', but which also defines invertebrates such as daphnids as outside its scope and definition of the types of animals requiring care. The PhD student described the way these laws define 'what is worthy of respect and care and consideration' as 'disappoint-ing', and indicated that he thought that the principle of 'reduce, refine and replace' should be applicable for experiments involving all animals. Without this, he felt that the law was drawing a very sharp line that allows you to 'do a lot of perverted and sick things to animals with-out anybody having any legal right to question your work'. He felt that 'there should be a basic respect for living organisms', but that this 'is just not there, and as researchers we are not trained to have that'. It would therefore be interesting to consider and experiment with how such training may be implemented in practice.

The sharp line drawn in the law on animal welfare, which disappoints the scientist here, is certainly not sharp within environmental ethics and its subfield of work on animal ethics and rights (e.g. see Garner 2005; Regan 1983; Ryder 2000; Singer 1975). Establishing any grounds for where such a line should be drawn is a matter for on-going debate. When the scientists consider how they should treat their test organ-isms, both in terms of the law and in terms of their own beliefs, they inevitably engage with this debate. Interestingly, when asked to reflect upon and discuss this question of the boundary of our moral commu-nity more deeply, all of the scientists made some reference to the issue of the size of an organism, indicating that size somehow played an intu-itive role in their definition of the boundary '*Daphnia* is easier than mice, I think it is just because it is smaller' (M). 'Size makes a differ-ence. There is a difference between working on a whale and an ant'

(LL). Size, however, is not a criterion that typically features in the reasoning of animal rights scholars and, indeed, when pushed to further elaborate and think through this line of argument, all of the scientists retreated from this as a valid criterion and sought other justifications for why certain animals may deserve enhanced levels of care and respect. These other justifications were typically connected to how closely the organism is related to us in an evolutionary sense, using the concept of higher levels of organisation or orders of organisms. 'I think it is because it (the *Daphnia*) is further from us, biologically, so we relate to it less. A mouse is a mammal like us ... For me it is sadder, for instance, to have chimps or monkeys used in experiments, maybe it is because I relate more to them ... I think in my mind it is because they are, in terms of consciousness, they are more evolved, than *Daphnia* for instance' (M).

The use of ecological concepts, such as evolution, to help structure an environmental ethics framework was also, arguably, what motivated the scientists to introduce Albert into the laboratory and claim that this represented a demonstration of care and respect for the daphnids. All of the other apparent demonstrations of care, such as gentle handling, etc. also had a strong scientific function. That is, it was also beneficial for the experiments that the organisms were treated with care in these cases. This was not so with Albert. There was no scientific value to his introduction at all. This act was, in fact, voluntarily and not at my request, highlighted to me on my very first day in the laboratory as a measure taken by the scientists to create a 'more ethical' situation. The scientists were aware that the use of Albert was difficult to justify from some perspectives: 'I think the poor *Daphnia* were probably shit-scared because you know the *Daphnia* can feel that the fish is there ... there is scientific evidence ... that the daphnids can sense predation' (PhD); 'I was putting them to their death by giving them to Albert ... and maybe they would perceive the predator and maybe it would be scary for them, I don't know'. Yet, when pressed, their justification of his introduction as an ethical act, an act of care and respect, was through the ecological concept of a food chain or a 'natural trophic cascade' (PhD). This idea was that this represented a 'more natural way of dying ... it was the thing that we could do that could bring them closer to nature' (M) and was therefore also somehow more ethical.

Such justifications and positions touch upon one of the other core debates that circles the question of the boundaries of our moral community in environmental ethics, that is whether the appropriate entities for environmental value are individuals or communities/systems (e.g. see Callicott 1989; Leopold 1968). While interesting discussions were had with the scientists about the value of knowledge generated through experiments performed on individual organisms under controlled laboratory conditions for understanding effects in complex dynamic natural systems, and how this should be handled in experimental design, data interpretation and result communication, this topic deserves more attention than is possible within this chapter. However, when the scientists justify the use of Albert as an ethical act, their belief in the value of a systems-based ecological food web can be seen to be overriding the perceived value of an individual daphnid's welfare (i.e. their fear, as suffering, is discounted). Another interesting element that touched on this debate about an individualistic versus a systemic approach to environmental value, was when the laboratory leader suggested that the instrumental use of individual daphnid's was less problematic than if whole populations of organisms had been used, and, indeed, that removing them from natural populations would have been even more problematic than producing them in the laboratory's own 'mother factories'. Additionally, the fact that they were clones was, half-jokingly, suggested as making it less problematic, as you could create many copies of the same individual.

Through elements such as their use of ecological concepts to structure their deliberations over questions of environmental ethics and their justification that the use of individual organisms in the laboratory helped contribute to a greater good of protecting natural systems more broadly, the scientists arguably demonstrated a tendency towards a more systems-based approach to articulating environmental value. That is, within their ecological world view, a higher value was seen to reside in populations and functioning ecological systems, rather than in individual organisms. The scientists were able to justify their instrumental use and treatment of individual daphnids as legitimate or acceptable because they perceived a higher value residing in larger systems.

Beyond environmental ethics

These represent just some of the ways in which central issues of debate within environmental ethics are encountered by ecotoxicologists on the laboratory floor and they will certainly represent fruitful areas for on-going investigation throughout the latter stages of the project. There were, however, also other interesting ethical issues related to their scientific practice that were encountered that were not specifically environmental. These included issues such as how research questions are chosen, how data is transformed through interpretation and statistical analysis, and how communication about research is conducted. There were also larger philosophical questions that the ecotoxicologists had to wrestle with when collecting their data, such as how to define when a daphnid was dead, as well as what counts as a life [e.g. in terms of distinguishing between an aborted foetus and 'just mush' (M)]. Given that mortality is a key endpoint in toxicology, both of these issues have a high level of significance for the process of data collection, the results that come from the studies, and their interpretation and effect on political decision-making. There were also additional ethical issues specific to working with GMOs, such as difficulties in gaining access to test materials without resorting to illegal means, and the often vicious and vindictive treatment of scientists publishing findings of adverse effects. Each of these issues represents potentially fruitful grounds for further investigation, engagement and the performance of ethical technology assessment.

Reflecting on my role and engagement in the laboratory

As stated earlier, the project that led to my laboratory engagement was funded by a programme with a belief that embedding ethicists in laboratory research could create an opportunity for developing more responsible research and innovation. From my own perspective, I had a dual intention in engaging in the laboratory. The first was to pursue a longer-standing interest in exploring how values, assumptions and beliefs influence and shape scientific knowledge. The second was to begin mapping how the issues of debate within the philosophy of environmental ethics were encountered and manifest in the practice of ecotoxicology. But what did the scientists think was my purpose in the laboratory? And how does this relate to the role I saw for myself and the role envisaged by my funding body?

In the interviews I conducted at the end of this first period of my engagement, the laboratory leader, the PhD and the Masters student all gave similar, and yet slightly different, descriptions of their perception of my role and interest in their work. What each shared was a perception that I wanted to better understand the realities of scientific research. For example:

> You wanted to learn how are things done in practice—so going from theory to practical work in the lab—and to learn and to see and to maybe have a perspective on what we did and how we decide things...and maybe an extended evaluation of say background values, ideas in the research, many questions that lie behind the research. So you would like to see it closely in order to learn and see in practice how things are done. (LL)

In this sense, there was a certain appreciation for my general interest in the practice of science and how values influence the choices that are made. However, whether my engagement had an ability to influence those choices, or created an opportunity for them to be made in a more 'responsible' or 'ethical' way is less clear. In my case, the hands-on approach that I specifically adopted could be interpreted as either facilitating or inhibiting my ability to affect research trajectories, methods and practices. All three of the scientists stated specifically that my hands-on approach meant that they did not feel 'objectified' during my period of engagement, which I was pleased to hear. For example:

> *Interviewer:* 'What if I had not been hands-on?'
> *M:* 'It would have been worse, more difficult, more work, and I think I would have had a greater feeling of being analysed maybe'.

Yet, it seems that the way I carried out the hands-on involvement possibly created a role for me that lacked an ability to influence research choices. For example:

> *PhD:* I felt that you had a genuine interest in the experiment itself, that you aided in the work, that you were in no way interfering with the processes that were going on ... I feel like you were putting yourself in a role where you were really assisting us, doing work that we had defined, according to ideas we had developed, and its ... yes, you were observing us and I felt that you were very conscious of what was happening in the lab, but you were in no way commenting or voicing your opinions or your values or evaluations so, I don't feel that you were, that your research into us was in any way evident or dominating, it was not in any way influencing the work that was going on ...
> *Interviewer:* 'You said I was assisting and aiding the work, can you say in what way?'
> *PhD:* 'Yes, you were working as a laboratory assistant. That's the function that you had. You were working as a laboratory assistant so you were doing tasks that had been defined, thereby easing the workload in the laboratory, and, on top of that, you contributed to the good spirits in the working process'.

Should I so wish, this may be something I could alter in the future, for example taking opportunities to express my opinions and evaluations

more clearly or more directly, in other words adopting a more normative stance in my engagement. This is, however, arguably not what is
typically envisaged in ethical technology assessment. That is, it is typically not seen as my role in the laboratory to evaluate the scientists and
give them my opinion on what the 'right', 'best' or 'ethical' thing to
do is, but, rather, to highlight where moral choices are being made, and
open for, or facilitating, their reflection on these matters. It could be
argued that such a process of opening for reflection was indeed taking
place.

> Your presence, I think, encouraged that, the reflection on these
> things, we had some talks about it and, because I think, for me, now
> it's more conscious. Before it was just a part of me, it was not com-
> pletely conscious why I was doing the choices I was doing, maybe
> I would not think about it beforehand, consciously. It would be a fac-
> tor that helps shape what I do but not so consciously, then now I have
> a bit more consciousness about it. I don't know if your presence in
> the lab changed my choices, but the fact that it made me more con-
> scious about it might have had an impact on my choices, I don't
> know if it had, but it might, I can say that . . . Someone questions why
> you are doing this and then you question, why I am doing this? And
> then you might think, oh no, maybe this is not the best way to do
> it . . . for instance if you ask me, why do you do that? It will make me
> think harder about it and question myself . . . and we were exchang-
> ing knowledge, because I was teaching you some techniques and you
> always learn when you teach. Because you have to think more about
> it and think in a way that the other person understands . . . (M)

Interestingly, when the scientist in this extended quote says that 'we
had some talks' about these 'things', these talks did not take place in formal interviews. Typically, they were conversations that took place while
working side by side at the laboratory bench. I worked most closely with
the Masters student, and, as indicated above, he and I were often the
ones responsible for carrying out the somewhat laborious daily tasks
in the maize experiment. This meant we spent a lot of time working
together one-on-one, and to pass the hours of somewhat repetitive task
work we often talked about issues such as choices in experimental design
and practice, environmental ethics, our values, spirituality and so on.
Of course, we also talked about music, *YouTube* videos and great beaches.
That is to say, the conversations did not always feel like research or
analysis, but they were building an understanding of each other and

encouraging reflexivity within both of our research processes. Additionally, it seems as though the hands-on process was specifically helpful in this, not just because of the time together it created, but also because the scientist specifically had to teach me techniques, and through my questioning to learn these techniques and practices he was forced to consider and reflect on them more deeply himself.

Outside of this sweet telling of engagement creating reflective opportunities and capacity, there are legitimate questions about the extent to which a researcher seeking to influence technology development through engaging in a laboratory can have an effect. This is not only owing to the fact that innovation is shaped by many social practices taking place beyond the laboratory floor (and we need to be careful not to forget or neglect these during the currently-in-vogue focus on integrated laboratory work), it is also owing to the extent to which groups are willing to grant outsiders access to all realms of their activity. For example, I was embraced in terms of observing and participating in the experimental laboratory work owing to reasons outlined earlier. However, I was often 'forgotten' to be included in meetings taking place around the planning of the work and/or the stage in which data collected was analysed and written into publications. This could simply be a feature that will improve with time and trust, or that could improve by my being more insistent on my interest in these stages of the process. To date, though, access to these stages of research in which important decisions concerning trajectories of future research and interpretations of significance are being made have proven the most difficult to access.

It is important to point out that this has not necessarily been an active blocking on the part of the scientists involved and may be related to my not being present in the laboratory 100% of the time. When I heard about such meetings or writing processes and asked to be involved there has been no question as to my participation, it is just that they don't think to include me from the outset, despite knowing my interest in researching and understanding the complete scientific process. I suspect that this difficulty possibly relates to the question of trading zones or, in other words, the question of whether I have something useful to offer at those stages that would make the scientists think to involve me. I am led to think this because, in one case, the leader of the laboratory actually consulted me rather heavily concerning one of the papers he was writing. This paper, however, was a response to a critique from a group of French scientists of one of his earlier works. The earlier work had been very central in a political decision to prohibit the cultivation of genetically-modified maize in Germany, and the French critique was questioning the quality of the scientific basis for the German decision. In this sense, the debate was focused on questions of choices in scientific research and the relationship between science and policy—areas where I did, in fact, have some expertise.[3] In this sense, I was invited to observe and participate, to some extent, in the writing process—in this case because I had some expertise to offer. In the analysis and writing of the scientific papers based on the experimental work of the laboratory, however, my inclusion in the process appears to be not as natural. That said, after requesting involvement in a recent effort to write a paper on the results of the maize experiment, and offering my native English skills as a relevant trade, I have not only been given access to the process, but also been recognised as a legitimate co-author on the paper (owing to my involvement in planning, laboratory practice and now writing). Working to improve my access to all the stages of knowledge generation, beyond data gathering through experimentation, is one of my aims for the latter years of my engagement, although it is likely to remain a challenge in short-term laboratory engagement exercises.

Conclusion

This chapter has described the beginning of an experimental engagement to explore the interface between environmental ethics and ecotoxicology motivated by a hope that connections can be made, and opportunities created, for the emergence of a more responsible ecological governance of technology development. Ecotoxicological science is invested with the power to shape the trajectories of new and emerging technologies, both directly through informing industry investment and formal processes of regulation, but also indirectly through shaping public opinion and governance more broadly. Understanding the ethical issues faced during the development of this science and exploring how they are negotiated in practice by the researchers involved, should be a crucial element of any ethical technology development.

Engaging with ecotoxicological science as part of any ethical technology development is particularly important to avoid a limited focus on just the stage of technology development itself. The significance of (eco)toxicological risk-based research for shaping our understanding, development, regulation and governance of new technologies makes this a particularly relevant and, until now, largely neglected site for such engagements. The relevance of laboratory engagements here may not necessarily be specifically in terms of whether such work can change the type of research being conducted or the research practices in operation (although this may, indeed, be possible); rather, it may be more about creating a better understanding of the realities of how this knowledge is generated (or constrained), including issues such as researcher motivations, the breadth of scope for different choices in the research, and the way in which values, politics and economics shape research choices and therefore our available knowledge about the risks technologies pose to human health and the environment. In this case, engagements on the laboratory floor may not be so much about changing the trajectory of the technology directly, but rather about building an understanding of the realities of risk-based research practice and thereby enhancing reflexivity about the role it currently plays in technology assessment processes.

The value for social science researchers in these endeavours should also not be underestimated. Typically, it is assumed that the aim of such laboratory engagements is to change scientific practice through enhancing the reflexivity of the scientists involved. My experience, however, has been that it is perhaps my own research practice and reflexivity that has been affected most by the process of engagement, and that it is some of the unintended and unanticipated outcomes that have proven to be the most interesting and fruitful. In this sense, it may be more beneficial to think of such engagements as collaborations or collective experimentations in knowledge generation, rather than a simple process of ethicists seeking to stimulate and steer responsible innovation. However, considering the engagement actors as 'ethicists' here may be problematic, given that ethicists are typically perceived (perhaps especially within natural science circles) as gloomy naysayers or barrier creators whose work prohibits progress. Indeed, after attending a five-day interdisciplinary workshop I held as a part of my project (in which the PhD student here also participated), one of the natural scientists declared that he had had a revelation. He arrived at the start of the workshop ready to defend his work and himself from what he perceived as the 'likely to be interfering' ethicists. After five days together though, he declared that he now realised that 'you are not ethicists, you are philosophers', and that this created a completely different space for our conversations and collaboration.

Engaging in laboratory research is one way in which we can pursue the goal of ethical technology assessment and responsible innovation. While it is worth considering and pursuing various methods to achieve our intended goals (e.g. building ethical reflexivity and social sensitivity through dedicated under- or postgraduate courses for scientists), specific value from engaging directly on the laboratory floor can be seen in the way in which this work is inevitably concrete, grounded and directly relevant for, and connected to, the scientists involved, rather than being abstract, vague and/or only distantly related to their daily practice. Experimentation with different modes for such engagement and on-going reflection about the aims and effects, opportunities and challenges of such exercises can only contribute to their further successful adaptation and evolution. It is therefore my hope that this chapter has been able to highlight ecotoxicology as a relevant site for ethical technology assessment and contributed to the on-going development of engagements taking place on the laboratory floor.

Notes

1. Although some authors draw a sharp distinction between environmental toxicology and ecotoxicology, where ecotoxicology is specifically informed by the science of ecology and seeks to perform more complex multi-species tests using exposure and a selection of test organisms based on realistic conditions (e.g. see Chapman 2002), in this chapter I adopt the more commonly used understanding of ecotoxicology as more broadly referring to all toxicological testing done with an intent to understand environmental effects.
2. Interestingly, the European Food Safety Authority (EFSA) recently released new guidelines for the environmental risk assessment of GMOs (EFSA 2010) in which they support a new approach to the selection of test organisms. This involves not just the use of standardised indicator species, but an identification and use of those with specific ecological significance. As a new approach, however, the practice of selecting and using ecologically relevant test organisms is yet to be implemented in the regulation of GMOs and ecotoxicological tests performed on the standard list of indicator species remains the norm.
3. It is worth noting that our discussions of this issue saw me not only give feedback to the laboratory leader on his own written response to the critique, but also lead to my own publication on the case (Wickson and Wynne 2012a) and a further piece on changes in the European regulation of biotechnology that were unearthed during research on the case (Wickson & Wynne 2012b), both of which the laboratory leader also gave me extensive feedback on. My early period of engagement in the ecotoxicology laboratory has therefore been interesting and fruitful in unintended ways.

References

Andow, D. A. and Hilbeck, A. (2004) 'Science-based Risk Assessment for Nontarget Effects of Transgenic Crops', *BioScience*, 54(7): 637–49.
Cairns Jr, J. (1995) 'The Genesis of Ecotoxicology', in Cairns Jr, J. and Niederlehner, B. R. (eds) *Ecological Toxicity Testing: Scale, Complexity and Relevance* (Boca Raton: Lewis Publishers).

Callicott, J. B. (1986) 'On the Intrinsic Value of Nonhuman Species', In: Norton, B.G. (ed.) *The Preservation of Species* (Princeton, NJ: Princeton University Press).

Callicott, J. B. (1989) *In Defense of the Land Ethic: Essays in Environmental Philosophy* (Albany: State University of New York Press).

Chapman, P. M. (2002) 'Integrating Toxicology and Ecology: Putting the 'Eco' Into Ecotoxicology', *Marine Pollution Bulletin*, 44: 7–15.

Demortain, D. (2011) *Scientists and the regulation of Risk: Standardising Control* (Cheltenham: Edward Elgar).

EFSA (2010) 'Guidance on the Environmental Risk Assessment of Genetically Modified Plants', *The EFSA Journal*, 8(11): 1–111.

Galison, P. (1997) *Image & Logic: A Material Culture of Microphysics* (Chicago: The University of Chicago Press).

Garner, R. (2005) *Animal Ethics* (Cambridge: Polity Press).

Hargrove, E. (2003) 'Weak Anthropocentric Intrinsic Value', in Light, A. and Rolston III, H. (eds) *Environmental Ethics: An Anthology* (Malden, MA: Blackwell Publishing).

Jasanoff, S. (1990) *The Fifth Branch: Science Advisors as Policymakers* (Cambridge, MA: Harvard University Press).

Lee, K. (2003) 'The Source and Locus of Intrinsic Value: A Re-examination', in Light, A. and Rolston III, H. (eds) *Environmental Ethics: An Anthology* (Malden, MA: Blackwell Publishing).

Leopold, A. (1968) *A Sand County Almanac* (New York: Oxford University Press).

Noer Lie, S.A. and Wickson, F. (2011) 'The Relational Ontology of Deep Ecology: A Dispositional Alternative to Intrinsic Value?', in Aaro, A. and Servan, J. (eds) *Environment, Embodiment & History* (Bergen: Hermes Text).

Norton, B. G. (1991) *Toward Unity Among Environmentalists* (New York: Oxford University Press).

Norton, B. G. (2003) 'Environmental ethics and weak anthropocentrism', in Light, A. and Rolston III, H. (eds) *Environmental Ethics: An Anthology* (Malden, MA: Blackwell Publishing).

Norwegian Government (2010) 'Animal Welfare Act', available at http://www.regjeringen.no/en/doc/laws/Acts/animal-welfare-act.html?id=571188 (accessed 22 June 12).

O'Neill, J. (2003) 'The varieties of intrinsic value', in Light, A. and Rolston III, H. (eds) *Environmental Ethics: An Anthology* (Malden, MA: Blackwell Publishing).

Regan, T. (1983) *The Case for Animal Rights* (Berkeley, CA: University of California Press).

Rolston III, H. (2003) 'Value in Nature and the Nature of Value', in Light, A. and Rolston III, H. (eds) *Environmental Ethics: An Anthology* (Malden, MA: Blackwell Publishing).

Ryder, R.D. (2000) *Animal Revolution: Changing Attitudes Towards Speciesism* (Oxford: Berg).

Singer, P. (1975) *Animal Liberation* (New York: Harper Collins Publishers).

Taylor, P. W. (1986) *Respect for Nature: A Theory of Environmental Ethics* (Princeton, NJ: Princeton University Press).

Taylor, P. W. (2003) 'The ethics of respect for nature', in Light, A. and Rolston III, H. (eds) *Environmental Ethics: An Anthology* (Malden, MA: Blackwell Publishing).

Wickson, F. and Wynne, B. (2012a) 'The Ethics of Science for Policy in the Environmental Governance of Biotechnology: MON810 Maize in Europe', *Ethics, Policy & Environment*, 15(3): 321–40.

Wickson, F. and Wynne, B. (2012b) 'The Anglerfish Deception: The Light of Proposed Reform in the Regulation of GM Crops Hides Underlying Problems in EU Science and Governance', *EMBO Reports*, 13(2): 100–5.

8
The Promises of Emerging Diagnostics: From Scientists' Visions to the Laboratory Bench and Back

Federica Lucivero

Introduction

Emerging technologies raise hopes and concerns about their potential consequences for societies. How will neurotechnologies affect our social standards of normality? How will new diagnostic technologies change the doctor–patient relationship? How should we deal with the privacy issues raised by biobanking? How will the data flow in brain–computer interface systems be protected? Asking these types of exploratory questions is important for modern societies in order to be prepared when these problems arise, and to guide citizens, policy-makers and technology developers to be more reflexive about their work (Grunwald 2010; Kass 2003). In this spirit, Western societies have, in the last decade, tried to investigate the so-called ethical, legal and social issues (ELSI) raised by new technologies.

However, this proved to be a difficult enterprise because emerging science and technology are still fluid and undefined objects. An ELSI analysis is based on *expectations* of what neurotechnologies, novel diagnostics, biobanking or brain–computer interfaces will be and do. In the last 20 years, a number of sociological studies have investigated how expectations surrounding emerging technologies play a strategic role in the dynamics of innovation (Brown et al. 2000; Hedgecoe and Martin 2003; Brown and Michael 2003; Selin 2007; van Lente, 1993; van Merkerk and Robinson 2006). It has emerged that expectations guide actors' decisions, legitimise their actions, attract interest or raise worries (Borup et al. 2006). For these reasons, expectations do not provide a stable ground for ethicists to reflect on emerging technologies. Furthermore, the socio-constructivist tradition has pointed out that new technologies can be expected to not simply have direct consequences on society, but, instead, to enter into a rich interplay of relationships with social actors (Bijker et al. 1987; Geels and Smits 2000; Tenner 1996).

For these reasons ELSI research has to be reflexive about its epistemological foundations (Grunwald 2010) and escape the danger of being too 'speculative', directing public attention towards implausible scenarios (Nordmann 2007; Nordmann and Rip 2009), or relying too much on the strategic promises of technology developers (Hedgecoe 2010). ELSI research needs to assure that it grounds on solid knowledge foundations.

In this chapter I will show how the laboratory floor offers such a rich ground and a stable springboard for ethicists to initiate an epistemologically sound reflection on the ethical, legal and social issues related to emerging technologies. I will discuss the case of a currently emerging biomedical technology for early diagnostics and health monitoring: Immunosignaturing (ImSg). Following a short note on my research design, I illustrate the type of information and considerations that I retrieved on the laboratory floor, pointing out for each of them their relevance to ELSI research. Laboratory studies provide an opportunity for ELSI studies to better articulate expectations surrounding emerging technologies and explore relevant ELSI dimensions. Later, I summarise the more general methodological learning obtained from this specific case.

Immunosignatures and the healthcare revolution

The Center for Innovations in Medicine (CIM) is one of ten research centres within the Biodesign Institute at Arizona State University devoted to the development of 'purposeful research to solve urgent societal challenges'. The Center's mission is to develop innovative research that attempts to transform our understanding of diseases. In many cases, innovation requires that we put aside what we think we know and start fresh.[1]

The CIM co-director asserts confidently that they aim to create 'a world without patients'. This is also the motto of the spinoff company that he has founded to manufacture microchips for research on 'immunosignatures'. This is how, in a video from 2010, the laboratory director describes this technology for 'harnessing the immune system's diagnostic power':

> One of the power projects in my Center is to have well people monitoring their health in a comprehensive way so that they can detect early any aberrations, anything that starts to go wrong with their health and they can do it early, and act early. We think that this is probably the most important thing that we do around the health ... in the United States also. [...] We realised that we have to develop a system that is cheap, very simple and very comprehensive, so that well people can use it all the time. There is a very powerful aspect of that, because it means that you are always normalising your health with respect to yourself. Right now in the biomarker world in medicine, as we live right now, we are normalising our markers to the whole population, generally not to ourselves, because we don't take them frequently enough to do that. So, this was our goal. We were trying to figure out how to do that and we knew that it had to be a *technological revolution* in order to be able to do these kinds of things. We finally came on this really *simple concept* and that was: you have millions, billions of antibodies in you and if those antibodies were always registering your health status and we had to look at that whole repertoire, and get a signature of your antibodies *in a simple way, we might be able to revolutionise diagnostics*. We used those same arrays with peptides that we were using to develop synthetic antibodies, and we said what happens if you put a drop of blood from some-body on there, and you wash it off and detect the antibodies? Well, it turns out that you get this signature; you get 10,000 spots light-ing up at different levels that basically fingerprints your antibodies. We said, maybe there is a change when something goes wrong. And sure enough we have tested over 20 different diseases now. Everyone shows *its own distinctive signature*. So we can normalise to yourself, and when something happens that signature changes. The beauty of it is that *measuring antibodies is so simple*. We can literally take less than a drop of blood, put it on a little filter paper, send it through the mail, even in Phoenix in the summer, take that filter paper and mea-sure the antibodies on it. The signature is just as good as if we had measured the blood directly. So what we envision now is a health monitoring system where people are regularly sending in [...] a little thing of saliva or a drop of blood, goes to a central place, they mon-itor, that information goes back to people and they can tell what

the health status is. And that's the big 'we are gonna go' that we are shooting for with this. What we show in this first publication is that this works very well to monitor infectious diseases, although we have other projects on cancer and Alzheimer that are going on.[2]

Immunosignatures are a 'technological revolution' able to transform diagnostics and improve the US healthcare system. It will relieve the national economy of healthcare costs—costs that are unsustainable given that medical technology, hygiene standards and healthier lifestyles have prolonged life expectancy and standards of medical care. Frequent personal monitoring is expected to prevent people from becoming chronically sick and thus permanently expensive for society. In this context, ImSg provides a system 'to have people monitoring their health in a comprehensive way and detect early any aberrations, anything that goes wrong with their system, and act early'. This is done in a 'simple' way: immune system activity can be disclosed by the detection of antibodies. The user will put a drop of blood on a piece of filter paper and send it by mail to a laboratory; the laboratory will analyse it; and this information is her/his personal health condition at a certain moment in time.

Although ImSg seems a very futuristic enterprise, the expectation of its success has resulted in several actions. For example, the Center's co-directors founded a spinoff company to support this project and to investigate the options to market it as a direct-to-consumer (DtC) test. Small kits have been assembled with filter paper, a lancet to prick the finger and instructions on how to send a biological sample to the CIM in order to have personal immunosignatures identified.[3]

These broad expectations present an emerging artefact that has a specific use in a social context and evident benefits. A reflection on the potential ELSI implications of emerging technologies aims at investigating the declared 'value' of this technology, its possible unwanted consequences and its desirability. But, as I will show in the following section, in order to explore these issues these broad scenarios should be abandoned, and expectations should be explored as they circulate within the laboratory.[4]

From the promise to the laboratory bench

I was involved with the CIM from November 2010 until January 2011 as 'embedded philosopher'. I was provided with a desk, introduced during a laboratory meeting to the whole group, and had access to all facilities, meetings and activities. I attended laboratory and project meetings twice a week, reviewed and discussed scientific papers and project proposals with researchers, and conducted participant observations during meetings and laboratory activities. I also held semi-structured interviews with several members of the research group: in some cases I had up to five interviews with the same person. Besides attending outreach events in which ImSg was presented to a lay audience, I also participated in informal activities (lunch conversations, the Christmas party). In addition, I presented my research twice at weekly laboratory meetings.

This interaction with researchers was aimed initially at understanding the science and technology of ImSg and its role in the project. When I became more acquainted with the scientific and technological aspects, it became clear that different applications were being explored by researchers. For this reason, I created additional instruments to engage CIM researchers in exploring their expectations of the context of use of these technologies, and organised two focus groups with researchers. The first focus group engaged eight participants around the topic of 'the challenges and promises of immunosignatures'. The goal of this discussion was to collect researchers' discourses on the feasibility and challenges of ImSg. The second focus group engaged 14 participants and focused on 'the applications and practices of immunosignatures out of the lab'. The goal of this second discussion was to collect researchers' ideas about potential applications and to invite them to articulate their descriptions of possible contexts of use.

Reconstructing the history of the concept

ImSg is presented as a 'technological revolution' that is founded on a 'simple concept': our immune system mirrors our health. Our immune system produces antibodies as a response to the presence of foreign bodies in our system. A read-out of the antibodies would thus provide information about the activity of the immune system and the health condition of the person who provided the sample.[5] ImSg provides this read-out together with an interpretation of it. Reconstructing the origin and history of this revolutionary 'simple concept' within the CIM is important in order to understand the features of these expectations.

According to the CIM co-director, the whole project started as a way to support another research project: the cancer vaccine. The researchers needed to find a way to validate the vaccine and demonstrate that it had, indeed, initiated a change in an organism. So, as a senior researcher explained, they started with the question: 'Can we make an array to measure 1000 *things* in the blood?'. They thought about measuring 1000 proteins as biomarkers of health conditions, and they wanted to use antibodies on an array to catch the presence of these proteins.

They soon realised that the different antibodies had recognisable binding patterns. Researchers wondered 'What happens if we put a mixture of the two antibodies on the array?'. They tried it, and could see a mixture of the two patterns on the array. This finding made them believe that they could retrieve this information from antibodies within the body. They tried it with human blood on the array and saw a pattern that they interpreted as a mixture of the patterns of the different antibodies present in the blood. Antibodies thus became the things in the blood that they wanted to measure, while the proteins (in fact, a form of simplified protein, 'peptides') became the 'catchers' on the array. 'This is what started the immunosignature array and we scaled it up to 10,000 peptides (simplified proteins)'. They ran samples from the same person before and after a vaccination, and saw big changes. In fact, these results showed that (i) it is possible to distinguish the signatures of different diseases in samples from the same individual, and (ii) the signatures of different individuals are dissimilar. A senior researcher who has been working at CIM for several years explains:

> The group, at that point, split up and this became its own project. Before, we thought that you had to scan every single protein in your body, but if we find a way to read this out we don't need to know every single protein, but your antibodies contain enough information.

The original idea has evolved in such a way that what is registered in ImSg is not the presence of certain proteins, but, instead, the presence of particular antibodies. These antibody patterns are now considered to have meaning for the health condition of the tested person.

The history of the immunosignature idea is described by researchers as a 'paradigm shift' in their understanding of the relationship between the detection platform and biological information: what was initially the platform (the antibodies) became the information being sought, and what was the information (the proteins) became the detection platform. The initial idea of ImSg has evolved into a new scientific and technological platform, while the function and role of this emerging technology is considered to be invariant. By reconstructing the history and evolution of ImSg, it is possible to differentiate projects and visions that are clustered and undistinguished in researchers' discourses. An ELSI reflection can therefore be more vision-specific and avoid implausible generalisations.

The 'paradigm shift' was accompanied by a second novelty: the use of random peptides (chains of amino acids) on the microarray. The current paradigm in immunology is that antibodies are antigen-specific, that is antibody x will bind to specific foreign body, antigen y. So, in order to 'catch' the antibodies in a biological sample, specific proteins should be used in the detection platform (a glass microarray). However, this is not the case on the CIM platform. Here, chains of amino acids (peptides) are randomly assembled and co-located on a glass array. These peptides are *randomly* generated in the sense that they present some of the 'bricks' (amino acids) that make up specific antigens, but these bricks are arranged erratically. This means that no one specific antigen is on the glass array. When researchers put a drop of human serum on the array they expect the antibodies in the serum to bind to the peptides on the glass plate. By expecting antibodies to bind to random chains of amino acids, researchers at CIM challenge the paradigm in immunology of antibodies as being antigen-specific.

After a researcher returned from a conference, he shared with his group the resistance of the scientific community to the group's use of 'randomly generated peptides'. His audience couldn't grasp how, on their array, they had chains of amino acids generated randomly by a computer, rather than using chains of amino acids corresponding to a specific pathogen. The need to overcome this resistance from the scientific community emerges in many conversations with CIM researchers.[6]

By pointing out these scientific controversies and uncertainties, it appears that the 'simple' concept presented in the broad scenario is, in reality, more complicated and controversial. This type of consideration warns ELSI reflection about the actual state of the art and scientific knowledge around a specific project.

Between metaphors and research practices

As the co-director explains in the aforementioned video, a 'fingerprint' of antibodies will provide truly personalised information because it allows people to 'normalize' their health with respect to themselves. An 'immunosignature' identifies an individual's health status at a particular point in time; whereas 'genetic signatures' provide stable, constant information about an individual's genetic make-up, immunosignatures are 'dynamic' in the sense that they reveal a snapshot of information about a continuously changing condition (the individual's immune response). An *immunosignature* is closer to a handwritten signature: authenticity is not granted by absolute permanence and identity, but by a *sameness* that is preserved over time. That sameness guarantees the link between a changing *token* and an immutable *type* or individual subject behind that. The ontology introduced by the concept of *immunosignature* differs from the ontological assumptions of the concept of 'genetic signature'. Genetic information offers information about an individual's genetic make-up that can lead to a deterministic acceptance of character and behavioural traits.

However, a closer analysis of scientists' research practices allows the ethicist to better qualify these broad statements about the personalized character of immunosignatures. I learnt that the similarity between today's or yesterday's patterns is made manifest by analysis of data collected on the glass plate. Samples of individuals with the same disease are expected to show a similar 'pattern'. So researchers are not interested primarily in *individual* signatures, but rather in *standard* signatures. It is important to have a baseline to describe when a signature is normal, when it is abnormal, and when it is going towards a pattern indicating cancer or Alzheimer's disease. Identifying such configurations requires huge statistical effort given that researchers are looking at differences both between people and between diseases. As one graduate student explained, they need to identify a 'standard normal signature', that is a reference line of ImSg, a range which gives an idea how normal individuals (free from any common chronic disease, irrespective of gender and age) respond to a particular peptide on an average.

So 'personalised signatures' might be a long-term goal, while standard normal signatures are the short-term goal most researchers are focusing on. Focusing only on the personalised character suggested by the signature metaphor is misleading for an ELSI reflection on such emerging technology.

What are 'immunosignatures'? Articulating objects and concepts

The 'immunosignature' is a concept coined by CIM. It is defined as the actual 'image' (or 'snapshot') of spots on the array obtained when the glass plate, with antibodies on it, is scanned. As a researcher explained to me:

> With traditional tests, you're only analysing the immune response to very defined things. The question addressed by IMS is not 'Is there a particular disease or infection going on?', but 'What is going on?'.

Thus, ImSg should be the technique, method or platform for identifying immunosignatures concerning a general health situation, rather than the detection of a specific antigen.

Random peptide microarrays are an 'unbiased' way to test antibody repertoires, such that researchers do not need to have a 'preconceived idea' of what peptide (and thereby pathogen) they should use to detect a specific antibody. In this way they can look at 'patterns' on the array without needing to know the pathogen. According to them, this is a good way to know whether there is any disease present. However, in some projects more specific information is relevant, and therefore a different platform will be used. For example, a group of researchers was working on a project on biosecurity aimed at testing soldiers for specific bio-threat agents. They had to select specific pathogens (e.g. anthrax or smallpox) and place them on the microarray. The same procedure was done by a researcher working on an infectious disease typically found in the southwestern USA (Cocci, or Valley Fever).

Therefore, researchers 'do immunosignatures' in at least two ways: (i) they use an array with 10,000 random peptides or (ii) they use arrays displaying only some specific epitopes of known pathogens. These two ways of doing ImSg address different clinical questions: 'What's going on?' versus 'Is there a particular infection or disease going on?'. These considerations are important to this analysis because the concept of an 'immunosignature' hides these differences between material practice and application. At least two technological platforms can be pointed out distinctively under the same 'immunosignature' label. In the following I will show that these platforms can be expected to be used in different contexts.

The possible applications of ImSg

In the broad scenario described by the CIM co-director (see the section *Immunosignatures and the healthcare revolution*), ImSg is a desirable innovation because it allows for a reduction in healthcare costs that are unsustainable for the US economy. ImSg enables a 'cheap', 'comprehensive' and 'regular' monitoring system for healthy people: 'well people can monitor their health in a comprehensive way and detect early any aberrations, anything that goes wrong with their system, and act early'. Frequent personal monitoring would allow doctors to intervene early in the case of any aberration and thus prevent people from becoming chronically sick and permanently expensive to society. However, on the basis of the previous analysis, we can ask *which* ImSg is expected to have this desirable consequence.

This CIM official discourse often focuses on the 'disruptive' character of their innovation. ImSg promises a revolutionary solution to healthcare problems. ImSg is not a diagnostic test for telling whether people *are* sick, but a system for pre-symptomatic monitoring of healthy people that tells you whether people are *in the process* of becoming sick. As such, the CIM directors and one graduate student are exploring possibilities for introducing ImSg to the market as an online DtC test. They have been exploring the business models of successful companies, such as 23andMe, that provide DtC genetic tests.[7] In doing this, they have to address the problem of the regulatory constraints that the Food and Drugs Administration[8] has placed on online sales of diagnostic genetic tests.[9] Companies like 23andMe have found creative solutions to these constraints, including offering these kits for 'personal genetic information' rather than 'diagnostics'. One possibility for marketing ImSg at this early stage would be to follow this model and offer 'raw data' to consumers about their immune system activity. The users would receive information about their immune system activity, without any interpretation of their meaning.

While some researchers address these kinds of questions, most CIM activities seem to focus on defining the 'patterns' for specific diseases: infectious diseases, influenza, diabetes, breast cancer and Alzheimer's disease. One graduate student collaborates with a medical doctor and expert on Valley Fever. She explained to me how the doctor has provided samples from patients who have been checked for the fungus that causes the disease. Her project is aimed at exploring the possibility of the early diagnosis of this infectious disease; preliminary results in this are promising. ImSg is therefore not only a 'disruptive' innovation for monitoring healthy people, but also provides a platform for more 'incremental' applications for diagnosing symptomatic diseases. The unifying vision of a system for monitoring the general health of asymptomatic people doesn't do justice to this variety of research.

Fleshing out possible context of use of ImSg

Distinguishing technological platforms and their potential applications is a first step in further exploring the different contexts of use of an emerging technology, such as ImSg. In the second focus group I explored this variety of potential applications. I asked participants to list and describe the short-term applications for ImSg. After some discussion among themselves, they agreed that it is more likely that, in the short term, the ImSg platform will be applied to enhance current diagnostic practices for specific chronic and infectious diseases, rather than being a tool to comprehensively monitor the health status of asymptomatic people.

One of the scenarios articulated during the focus group focused on ImSg as a platform for the diagnostics of a specific chronic or infectious disease. In the case of Valley Fever, researchers drew parallels between this potential system and the current US procedure for testing for strep throat. A symptomatic patient will be asked by their medical practitioner to perform the test through the collection of a drop of blood. The test will then have to be sent to the laboratory for analysis and, in the case of positive results, the patient needs to go back to the primary care doctor in order to receive treatment.

A second scenario elaborates a general vision of ImSg as a tool for the comprehensive monitoring of healthy people. Asymptomatic, healthy people would regularly send a biological sample to a central laboratory and receive their results by email. An online system could be in place in which people could log in, order the testing kit and receive it at home. After putting the drop on the filter paper, the sample would be shipped, probably via a post office so that an officer could write down the content and the reason that they are sending a biological sample. A central laboratory could analyse the sample and upload the result to the online platform so that users could have online access to their immunosignatures. This scenario presents two phases. In the short-term, while research on the correlation between immunosignatures and health status is still on-going, this system is expected to provide information about a person's immune response, without providing any interpretation. This information could be shared openly and 'people' with computational skills might then be able to find some correlations. In the long-term, it could be uploaded to some platform (such as *Google Health*) where people could collect and manage their own personal health records.

These scenarios, discussed by researchers, are interesting because they flesh out different ways of envisioning the future of the healthcare system, and the place of ImSg in it. They illustrate how an apparently unitary expectation of a technology opens up very different contexts, with different kinds of questions. The vision of ImSg as a 'comprehensive monitoring system for healthy people' thus competes with a vision of ImSg as a 'diagnostic tool' for specific health conditions in symptomatic patients.

Different contexts and different values

As explained, when we talk about ImSg we can distinguish at least two technological platforms with different applications and contexts of use. The first platform is expected to be used in the context of comprehensive pre-symptomatic health monitoring, where there is no suspicion of a specific syndrome. The second ImSg platform, which addresses specific diseases or infections, is expected to be used in a clinical context in which a diagnostic decision has to be made. These ImSg platforms are not only diverse in their techno-scientific components and in the clinical (or, more generally, social) context in which they are expected to operate. They also promote different values.

The *DIY online comprehensive monitoring system* is expected to provide online customers with information about their immune system activity. In the short-term vision this information is provided to clients, but no interpretation of its meaning is given. Based on the business model used by other DtC test providers, it is assumed that individuals own such information about themselves, and that what they do with this information is up to them. In this 'patient-centred' healthcare vision, the liberty of individuals and their self-determination are therefore central values. Individuals will not be dependent on doctors, and they will be able to take care of their health in an autonomous manner; ImSg is seen as empowering individuals in this respect. Interestingly, these 'libertarian' values of freedom—based on the concept of personal property—are combined in researchers' scripts with 'communitarian' values.[10] In fact, CIM researchers point explicitly to e-healthcare platforms in which users share their personal health information and data with other users in order to maximise the availability of interpretation tools.[11] This sort of open-source tool doesn't aim at accurate interpretation, but its value would be to provide a platform for healthy people to share information and knowledge concerning the relationship between data on immune system activity and the meaning of such data in terms of condition of health. The assumption is that people, while owning their personal information, need to share it with others in order to make sense of it. The moral connotation of this envisioned technology is therefore an interesting combination of libertarian ideas of freedom and autonomy together with a communitarian ideal of 'sharing'. Let's compare the moral meaning of this vision with the others.

The vision of ImSg as a *tool for the diagnosis of a specific disease* enables a system in which doctors play a central role. Here, ImSg is considered to be an empowering tool for the doctor, rather than the individual user. In this context, ImSg is a good test not if it promotes individual autonomy or freedom, but if it promotes accuracy and salience within clinical practice.

By referring to the economic value of a health monitoring system for asymptomatic people, other moral connotations that are inscribed in co-existing technological platforms remain unarticulated. Articulation of these diverse moral connotations shows that multiple artefacts, applications and values should be distinguished before the question of the ELSI of ImSg can be addressed.

The learnt lesson from the case study

ELSI research has to deal with expectations surrounding an emerging artefact. However, the descriptions of an emerging artefact aimed at a general public are often simplified abstractions. Whereas these public expectations describe the future artefact as an accomplished cold device, the laboratory offers interesting access to expectations in which the history and making of the artefact is still visible. Here, researchers' choices, uncertainties, challenges, controversies and doubts, concealed in public expectations, become visible. Furthermore, in research practice, alternative design choices co-exist and are discussed, and some design aspects too complicated to explain to a broader audience emerge. This laboratory-based analysis of expectations makes ELSI reflection less speculative and more attentive towards specific aspects of currently emerging technologies.

Through the example of immunosignatures, this chapter has illustrated an approach based on two strategies: situating and thickening. I elicited the expectations in the situated practice of the artefact making, where expectations were embodied in experimental setups and research tasks. New aspects of the expected artefact, collected during the situating exercise, were added to the original expectations, while others were removed, resulting in 'thickened' descriptions of the artefact-to-be. The reconstruction of the origin and history of the ImSg science and technology has pointed out the novelty of the concept of 'immunosignatures'. By using antibodies as a source of information about the condition of health, and randomly assembled peptides to catch them, ImSg challenges two 'paradigms' in the microarray and immunology world. For this reason, according to researchers, it triggers the disbelief of their peers. By looking at the research practices, the metaphor of immunosignature as 'personalized' information can be rephrased within the actual search for a standardized signature. My analysis of expectations on the laboratory floors has also disclosed the different ways that ImSg is done in research practice, and suggested that alternative technologies, with different expected uses, co-exist behind the idea of ImSg. When situated in the actual scientific practice of 'doing immunosignatures', the promise of ImSg acquires new dimensions. In particular, new scenarios emerge, such as the use of immunosignatures as diagnostic tools for specific diseases. Together with ImSg researchers I have articulated scenarios describing the potential contexts of the use of such technologies. On closer analysis, I pointed out that these scenarios reveal the promotion of different values to the ones mobilised in broad promises and expectations on ImSg (as the example of the video suggests).

On the laboratory floor, ethicists can articulate expectations surrounding emerging technologies. In fact, in their daily research, researchers deal with the problems, challenges and uncertainties involved in single sub-projects and initial hype is reduced to more modest expectations. In this way, ethicists can make explicit the meaning of metaphors and point out technical conditions that change the discourse surrounding an emerging technology (as in the case of personalized versus standardized signature). Furthermore, the laboratory provides a site to focus on the materiality of the different components and tools, and the research activities in which researchers are involved. In this way, ethicists can point out alternative technological platforms that might be hidden under the same label. Such a distinction is important because these technological platforms might suggest different applications. The ethicist can contribute in the articulation of the scenarios describing the diverse context of use of such technologies and point out the conflicting values promoted by these alternative technological platforms.

This type of analysis of expectations surrounding emerging technologies on the laboratory floor allows ELSI research to address the need of being less speculative and more critical towards strategic technological promises. Furthermore, when appropriately fed back into the laboratory floor, this type of analysis can guide scientists in making decisions about desirable innovation paths.

Notes

1. From the Centre for Innovation in Medicine website: http://www.biodesign. asu.edu/research/research-centers/innovations-in-medicine
2. Video available at http://vimeo.com/12370576. Transcript and *emphasis* are mine.
3. This kit had only been used for research purposes at the time of my fieldwork.
4. These ideas are inspired by Van Lente (1993). In a pioneering study on technological promises, Van Lente shows that expectation statements circulate at different levels and stages of the technological innovation process with different functions. Expectations at the 'macro-level' are general *promises* and scenarios on the technology used by technology developers to legitimise an emerging technology. At a 'micro-level', expectations circulate among experts in a laboratory in which they are expressed as quantitative *specifications* about future artefacts. These 'search' expectations guide decision-making and agenda-setting with a research group. At an intermediate (meso-)level, expectations are qualitative statements about the *functions* that the technology will presumably fulfil. They are general statements used by actors outside their own research context to position themselves within a certain field, but also to show the opportunities of a field and guide decisions for funding allocation. The boundaries between the macro-, meso- and micro-levels are fuzzy. Van Lente stresses that statements, agendas and actions among different levels are 'nested': the acceptance of a broader promise requires the acceptance of other expectations associated with it.
5. 'Our basic premise is that the antibody profile from an individual reflects their health status. If this profile can be displayed on a sufficiently complex array, the particular responses to chronic diseases will be apparent' (Stafford et al. 2012).
6. However, sometimes this controversy lurks in the discourse of researchers working at CIM. For example, CIM researchers talk about 'real' peptides in reference to known sequences that are recognised as being the target of specific types of antibodies. They call 'artificial' the peptides randomly assembled that they use on the ImSg array. 'In science nobody believes you if you don't show that actual stuff in the body. Here you show random stuff, artificial, harder to convince scientific society who only believes in real stuff" or 'this is more real...random peptides is a sort of artificial'. This dichotomy 'real versus artificial' suggests a hierarchy in the ontology of the researchers. The peptides synthesised in the laboratory from random amino acid sequences are less valuable, or less trustworthy, than ones traceable in nature. On one hand, researchers think that the ImSg concept does make sense and that the 'conceptual hurdle' for the scientific community can be overcome by showing data that support this unconventional view of how the immune system works, hence publishing, showing evidence, producing results, winning grants, etc. On the other hand, by speaking of 'real' or 'artificial' peptides, the researchers embed in their daily language the scepticism

of the scientific community. There is more. This distinction between 'real' and 'artificial' epitopes also discloses a difference in what the concept of 'immunosignature' means.

7. See the website where this direct-to-consumer genetic profile is sold: https://www.23andme.com/

8. The Food and Drug Administration is the US agency responsible for protecting and promoting public health through the regulation and supervision of, among other things, pharmaceutical drugs, medical devices and vaccines. See http://www.fda.gov/

9. See Little (2006).

10. For a short introduction to libertarian and communitarian values see Swift (2001).

11. See, for example, www.patientslikeme.com

References

Bijker, W.E., Hughes, T.P., and Pinch, T.J. (1987) *The Social Construction of Technological Systems: New Directions in the Sociology and History of Technology* (Cambridge, MA: MIT Press).

Borup, M., Brown, N., Konrad, K., and Van Lente, H. (2006) 'The Sociology of Expectations in Science and Technology', *Technology Analysis & Strategic Management*, 18(3-4): 285–98.

Brown, N., Rappert, B., and Webster, A. (2000) *Contested Futures: A Sociology of Prospective Techno-Science* (Aldershot: Ashgate).

Brown, N. and Michael, M. (2003) 'A Sociology of Expectations: Retrospecting Prospects and Prospecting Retrospects', *Technology Analysis & Strategic Management*, 15(1): 3–18.

Geels, F.W. and Smits, W.A. (2000) 'Failed Technology Futures: Pitfalls and Lessons from a Historical Survey', *Futures*, 32(9-10): 867–885.

Grunwald, A. (2010) 'From Speculative Nanoethics to Explorative Philosophy of Nanotechnology', *NanoEthics*, 4(2), 91–101.

Hedgecoe, A. and Martin, P. (2003) 'The Drugs Don't Work: Expectations and the Shaping of Pharmacogenetics', *Social Studies of Science*, 33(3): 327–64.

Hedgecoe, A. (2010) 'Bioethics and the Reinforcement of Socio-technical Expectations', *Social Studies of Science*, 40(2): 163–86.

Kass, L.R. (2003) 'Ageless Bodies, Happy Souls: Biotechnology and the Pursuit of Perfection', *New Atlantis*, Spring (1): 9–28.

Little, S. (2006) 'FDA Regulations and Novel Molecular Diagnostic Tests', *Clinical Laboratory International*, 7: 48–9.

Nordmann, A. and Rip, A. (2009) 'Mind the Gap Revisited', *Nature Nanotechnology*, 4(5): 273–4.

Nordmann, A. (2007) 'If and Then: A Critique of Speculative NanoEthics', *NanoEthics*, 1(1): 31–46.

Selin, C. (2007) 'Expectations and the Emergence of Nanotechnology', *Science, Technology & Human Values*, 32(2): 196–220.

Stafford, P., Halperin, R., Legutki, J. B., Magee, D. M., Galgiani, J., and Johnston, S. A. (2012) 'Physical Characterization of the "Immunosignaturing Effect"', *Molecular & Cellular Proteomics*, 11(4): M111.011593.

Swift, A. (2001) *Political Philosophy: A Beginners' Guide for Students and Statesmen* (Oxford: Polity).

Tenner, E. (1996) *Why Things Bite Back: Technology and the Revenge of Unintended Consequences* (New York: Knopf).

Van Lente, H. (1993) *Promising Technology: The Dynamics of Expectations in Technological Developments* (Enschede: Universiteit Twente).

Van Merkerk, R. and Robinson, D. (2006) 'Characterizing the Emergence of a Technological Field: Expectations, Agendas and Networks in Lab-on-a-chip Technologies', *Technology Analysis & Strategic Management*, 18(3/4): 411–28.

9
Dramatic Rehearsal on the Societal Embedding of the Lithium Chip

Lotte Krabbenborg

Introduction

There is a certain immediacy, and thus attractiveness, in the idea of encountering scientists in their natural habitat, on the laboratory floor. As ethnographers know, however, a lot of the time nothing much happens, other than the daily practices and their exigencies. The ethicist or social scientist might become more active, ask questions and point out (ethical) issues. But, even then, these only lead to minor modifications (Schuurbiers and Fisher 2009). One key point is that scientists act on many more 'floors' than the laboratory floor, and encounter different groups of actors, particularly when scientists are involved in the development of new technologies (Sorensen and Levold 1992). What happens on the different 'floors', for example in policy-making and funding, shapes what (can) happen on the laboratory floor, and how new technology and innovation will materialise in society (Rip et al. 1995; Robinson 2010).

In this chapter, I start from the other side, the many 'floors', and how these are linked to a particular newly emerging technology in the research and development (R&D) phase, that of lab-on-a-chip technology to stimulate point-of-care testing. I single out the idea that interactions between groups of actors operating on different 'floors' can take the form of 'dramatic rehearsal'. I borrow the notion of dramatic rehearsal from John Dewey (Dewey 1957). For Dewey, a dramatic rehearsal refers to a specific, ideal mode of deliberation required when people find themselves in indeterminate situations: situations in which it is not clear how to act, what to value or which ends to pursue. People feel blocked in their activities because existing routines, norms, values, roles and responsibilities are destabilised, for example owing to novelties introduced by new developments in science and technology, new ones are not yet in place. A dramatic rehearsal is then 'work of discovery' (Dewey 1957), an attempt to find out, by inquiry, imagination and experimentation, what is at stake, which ends to pursue and what to value.

Whereas for Dewey it is the existential experience of indeterminacy which leads to a dramatic rehearsal, in this chapter I will show how social scientists or ethicists can create an intervention during the R&D phase of newly emerging technologies that aims to *anticipate* indeterminate situations in the making and stimulate dramatic rehearsal. In order to create such an intervention, the role of a social scientist or ethicist must shift from the Socratic model that permeates many of the ethics on the laboratory floor studies, in particular the midstream modulation approach (Conley 2011; Schuurbiers and Fisher 2009) to that of a mediator who can be somewhat pro-active because of her/his own insight and preparatory work.[1]

I will describe how, using requirements from Constructive Technology Assessment (CTA) on emerging technologies, I designed and orchestrated a dramatic rehearsal around changing (societal) values, roles and responsibilities occasioned by the lithium chip. The process and outcome of the workshop will be assessed as to whether and how participants inquired, articulated and anticipated emerging indeterminacies in relation to the development and embedding of the lithium chip, and point-of-care-testing in healthcare more generally.

Dramatic rehearsal as a soft intervention

Social scientists or ethicists who want to create an intervention during the R&D phase of newly emerging technologies with the aim of anticipating emerging indeterminacies in relation to changing societal roles, values and responsibilities are facing challenges. Newly emerging technologies are still evolving. Which promises are to be taken up and how novelties will materialise in society, and what societal effects might be is not clear (Garud and Ahlstrom 1997; Robinson 2010). Where to start, who should be involved and how to design such an intervention? And why should a social scientist or ethicist organise such an intervention?

Is it not something that actors responsible for the development and embedding of new technologies should do themselves?

Although newly emerging technologies are characterised by a high degree of uncertainty, ambiguity and unpredictability, the first contours of destabilisation of existing societal practices do emerge. These contours are created by activities, considerations and choices that the groups of actors involved, such as scientists, industrialists, funders, policy-makers, insurance companies and civil society organisations, make. Activities of groups of actors set enabling and constraining conditions for how a new technology can materialise in society and which values, norms, roles and responsibilities might become dominant, and which ones will be silenced or remain under-developed.

Groups of actors involved could share and discuss their activities, inquire how these are entangled and explore what new societal practices are in the making. And, in fact, actors do point out possible societal effects of newly emerging technologies. However, owing to the way newly emerging science and technology is enacted in our society, it is relatively easy for actors involved to put aside these questions during the early stages of a technology development because these are not part of their (institutionalized) tasks and mandates. In our differentiated society, a concentric approach to technology development and societal embedding is dominant: addressing hopeful promises and functional issues first, and sequentially addressing broader aspects like regulations and societal embedding (Deuten et al. 1997). Thereby, owing to the fact that tasks and mandates to develop and embed new technologies are distributed,[2] groups of actors work in relatively separate (but not independent) worlds, which makes it difficult for actors to interact and become acquainted with one another's activities, perspectives and values. In the current division of labour, one can be a good scientist and make progress in science when one writes papers about the clinical functioning of lab-on-a-chip technology and never actually meet people from other worlds, such as policy-makers or representatives from funding agencies or civil society organisations.

CTA of emerging technologies has shown that social scientists or ethicists can take the role of a mediator, 'bridge' the separate worlds by organising dedicated 'bridging events' and articulate a diagnosis of an indeterminate situation (Van Merkerk 2007; Robinson 2010).[3] A social scientist or ethicist has a different position within an innovation trajectory. He or she can move around in the different worlds and observe, ask questions and study what is happening. Therefore, he or she can see things and point out issues, such as emerging patterns in the co-evolution of science, technology and society, that are more difficult for actors to see when they are operating in a particular domain and subjected to realising their daily tasks and mandates properly. A 'bridging event' can provide an opportunity for actors to freely explore emerging patterns, articulate problematic issues and develop strategies for how to deal with it.

A central tenet in the objectives and activities of the current CTA programme, especially how CTA has evolved over the past few years within the TA Nanoned programme (2004–2010), is that it primarily organized dedicated 'bridging events' that were focused on how to realize a 'better technology', for example by organising dedicated events to bridge the gap between innovation and ELSA[4] (cf. Te Kulve 2011 on institutional entrepreneurship and cooperation across the chain in the food packaging sector, and Parandian 2012 on strategies to break through waiting games in the domain of organic large area electronics). What has received less attention up to now in the CTA programme is the development of requirements to design and orchestrate dedicated interaction events that stimulate inquiry and articulation regarding 'better society', for example what societal practices are in the making as a result of new technology developments and how to value that (cf. Boenink, et al. 2010; Swierstra and Waelbers 2010).[5] To organise such an intervention, and develop requirements for who should be involved and what kind of interaction and participation to stimulate, the current CTA programme on emerging technologies can be complemented with Dewey's reflective inquiry approach to morality.

The point of departure for Dewey's reflective inquiry resembles the point of departure for CTA activities, namely uncertainty, unpredictability and ambiguity in current activities. In line with the CTA programme's attempt to design dedicated interaction workshops to explore uncertainties and articulate problematic issues, Dewey also proposed dedicated forms of interaction and participation, which he called dramatic rehearsal, to elucidate what is at stake. However, for Dewey, the spotlight in dramatic rehearsal is not on developing strategies for a better technology development, which is the central focus in current 'bridging events' of CTA, but Dewey's spotlight in a dramatic rehearsal is on exploring and defining the desirability of particular societal practices that are in the making. To phrase it in Dewey's terms: 'what is at stake [in a dramatic rehearsal] is what kind of person one is to become, what sort of self is in the making, what kind of world is in the making' (Dewey 1957). I will call the addition of dramatic rehearsal in CTA interaction workshops CTA+.

Dewey on dramatic rehearsal

Deliberation is not to calculate future happenings but to appraise present proposed actions (..) by their tendency to produce certain consequences (Dewey 1957, p. 193)

Dewey developed his work on dramatic rehearsal in the first half of the twentieth century with the aim of equipping citizens to deal in a more productive way with the rapid social, economic, demographic, political and technological changes they were facing, which led to the rise of urban–industrial societies in the USA. He noticed that new scientific and technological developments, such as public transport, the daily press and the radio, destabilised existing ways of living and associating (Dijstelbloem 2007), but that citizens were unable to interpret, value and judge these new developments because they inherited moral frameworks (i.e. values, norms, routines, perspectives) that stemmed from a pre-industrial era. With the development of the telephone and radio it became possible, for example, for people to maintain long distance relations without actually meeting one another. But in everyday interpretative frameworks, small-scale communities and face-to-face contacts remained the main point of reference. Dewey diagnosed that this led to 'uneasy equilibriums' between the old and the new (Dewey 1920, p. XXXII). His challenge was to reconstruct philosophy (that he perceived as a set of attitudes and beliefs) in such a way that it could serve as a guide for people living in a continuously changing and contingent environment (Dewey 1920). Dewey saw change, contingency, heterogeneity and unpredictability as characteristic of the human condition on earth. He noted that science and technology functioned as catalysts of change because it introduced novelties in existing frameworks that destabilise routines, values, norms and perspectives. To handle change, contingency and unpredictability more productively, Dewey argued that humans should see their interpretative frameworks as flexible and temporary. Dewey proposed to abandon the idea, dominant since Plato, that morality is something fixed and given prior to a particular situation. He perceived morality not as something that is imposed from the outside onto a situation. There is no special moral realm. Instead, Dewey understands morality as something that has to be *searched* for and *discovered* in the experience of everyday life. When people experience uncertainty about what to do, what to value and how to act, such a situation is a *moral* situation according to Dewey. Such uncertain situations are in need of deliberative reflective inquiry (i.e. dramatic rehearsal) to discover and determine, by means of deliberation, what to value, which ends to pursue and how to achieve them (Welchmann 1995).

A dramatic rehearsal refers to specific modes of deliberation and participation to inquire into indeterminacies (see Box 9.1).

Box 9.1 Interactive process of dramatic rehearsal

A dramatic rehearsal or reflective inquiry can be broken down into four subsequent phases (Dewey 1938, pp. 107–112; Hildebrand 2008, pp. 53–56). After the existential experience and recognition of an indeterminate situation (Dewey 1938), the first phase is to transform an indeterminate situation into a *problematic* situation by means of an inquiry and articulation of problems. Interactions take the form of sharing experiences, doubts and difficulties. Dewey emphasises that problems do not exist prior to an inquiry. In judging *that* it is a problem, we judge *how* it is, we define it (Hildebrand 2008). The second phase consists of formulating hypotheses about possible solutions to deal with the problems. Interactions take the form of forecasting, back-casting, challenging standard repertoires and imagination of possible consequences of executing a particular line of action. The third phase is a rehearsal, in interaction and imagination, of possible solutions. In this phase, an estimate of possible consequences is made for those who are involved. The final phase of a dramatic rehearsal is the experimental testing (in real life) of the hypothesis that emerged as best solution. Dewey stresses that the best solution should incorporate as many issues as possible that were discovered during the process (Dewey 1957). Dewey emphasises that interaction in a dramatic rehearsal should take the form of interaction and participation in the light of the unknown. What to value, how to act and which ends to pursue 'emerges' through interaction. A dramatic rehearsal thus requires that participants dare and are willing to perceive their values, norms and ideas as contingent and provisional, and to search for and develop new (or adapt existing) ones in relation to new situations.

CTA+ workshops to anticipate the co-evolution of science, technology and morality during the R&D phase of newly emerging technologies

The point of dramatic rehearsal (for Dewey, for me) is that people prepare for a future situation, as in rehearsing a play. The consequences and meaning of present proposed actions become visible because participants ('characters') (inter)act, and share their experiences, dilemmas and desires. To phrase it in terms of Dewey scholar Fesmire: 'To deliberate is to co-author a dramatic story with environing conditions in community with others' (Fesmire 2003, p. 78).

Dewey developed his work in an era that is different to ours. In order to maintain the core of Dewey's thought on dramatic rehearsal for the design and orchestration of a CTA+ intervention, I have to deviate from some of Dewey's assumptions and develop requirements that fit better with the socio-technical dynamics of the twenty-first century.

For Dewey a dramatic rehearsal is occasioned by actual consequences of actions that lead to the experience of disruption. People are affected by the consequences, and are able (at least to a certain extent) to recognise this and articulate how they are blocked in their daily activities. According to Dewey the experience of disruption forms the incentive for people to organise themselves and start a dramatic rehearsal because they have an urge to solve their problems in order to continue their daily practice.

Because science and technology is still evolving during the R&D phase, possible consequences for society are not clear. As such, concrete disruptions of existing (societal) norms, values, roles and relations are not experienced yet. However, first contours of destabilizations do emerge as a result of activities of different groups of actors. To stimulate dramatic rehearsal in a CTA+ workshop, the organizer can articulate these destabilizations by moving around in the different 'worlds' and develop a diagnosis.

To stimulate an inquiry into the indeterminacies, and articulation of what is at stake, the organizer of a CTA+ intervention can write her/his diagnosis down in a techno-moral future scenario. Such a scenario starts with a diagnosis of present uncertainties and ambiguities (in this case, the destabilisations of existing societal practices). Based on this diagnosis, the organiser can develop a narrative (controlled speculation; Te Kulve 2011) that shows (with action–reaction patterns) possible consequences and meaning of present proposed actions. Such a scenario can be seen as a virtual dramatic rehearsal. Those who are involved see themselves embedded in a broader development and the narrative shows how possible actions, interactions and repercussions might play out and lead to particular societal practices in which some norms, values, roles and responsibilities can be more easily pursued than others.

During the workshop, a scenario can function as a platform for actors (Te Kulve 2011) to articulate in interaction problematic issues, and develop and rehearse new or adapted norms, values, roles and responsibilities to address these issues. Participants can, of course, modify and add to the scenario.

Dewey is focused on individuals (citizens) and assumes that as soon as people start a dramatic rehearsal, they start to talk and question one another because they are curious to hear about one another's viewpoints and activities, and have a desire to acquire a better understanding of the complexities. This assumption is naive because, in practice, these forms of interactions are time consuming, and people are selective and make trade-offs about what they can gain from particular forms of interactions with a heterogeneous set of people. Analyses of CTA workshops have shown that interactivity between different groups of actors is not self-evident (Parandian, 2012). Enactors often position themselves as insiders who know much more about the technology and it's embedding in society. Actors who hesitate, articulate doubts or raise questions with regard to the promises the technology is expected to bring, are positioned as blind and irrational and thus not deserving of serious consideration. Actors not involved in the actual development of technology are labelled as outsiders, and even 'opponents'. (Rip and Talma 1998). These actors may not consider themselves as excluded, but enactors, nevertheless, define them as such.

To stimulate interaction and participation as dramatic rehearsal, the workshop organiser must meet orchestration requirements and emphasise that each participant participates as a *stake*holder (and not in the form of insider–outsider or expert-lay). Even if there is disparity between participants, for example with regard to (scientific) knowledge, in the aim of identifying problematic issues and preparing for a future situation, participants are symmetrical in the sense that each one is implied, although in different ways, in the emerging indeterminate situation, and each has particular views, experiences and knowledge with regard to what is at stake. By sharing these (so others can react) and by hearing others, understanding of the emerging new situation can be generated, and problematic issues can be uncovered in their full scope. A moderator can try to secure interactivity by inviting participants to share their experiences and dilemmas and stimulate them to actively question one another, listen to understand, and develop and rehearse possible lines of action to deal with problematic issues.

I will illustrate, in the next section, the content and process of a CTA+ workshop I organised around changing societal values, norms, roles and responsibilities in relation to the lithium chip, a nano-enabled device

to stimulate point-of-care testing. I will start with the emerging indeterminacy I identified by moving around in the worlds of lab-on-a-chip
technology and healthcare.

The lithium chip and its indeterminacies with regard to societal embedding

Lab-on-a-chip technology can be applied for different purposes. It's
possible use in measuring levels of electrolytes in the blood was explored
to create diagnostic sensors, in particular to measure lithium. This
particular case led to an opportunity to create a spin-off company,
which is now developing and testing such a device: the lithium
chip. The idea is to create point-of-care testing (POCT) for people who
suffer from a bipolar disorder and use lithium as medicine. POCT
refers to medical testing at any point, without the interference of
a (physical) laboratory. This point can be a patient's home, at the
bedside of patients or in a professional setting. POCT provides immediate results. Currently, patients with a bipolar disorder who use lithium
as medicine have to go to the hospital four or five times a year to
check their lithium level. The developers of the lithium chip see a
bright future where people can check their lithium level 'anytime,
anywhere' without the interference of a laboratory (www.medimate.
com). With just one drop of blood, the lithium chip will immediately display whether the lithium level of a patient is still within
the permissible range. Patients can act upon these results if they
think that it is necessary. According to the developers of the lithium
chip, quality of life will improve because patients will become less
dependent on professional healthcare and become more in control
of their own bodies. The lithium chip is still in the clinical testing
phase, but it is expected to be used in the near future. However, by
moving around in the different worlds, I noticed that those who are
involved—psychiatrists, patients, insurance companies, patient organisation and the technology developers—have different ideas about how
to introduce and use the chip, and have different perspectives on what
is at stake.

Existing roles, responsibilities and notions of good care will destabilise, but it is not yet clear in what way. Psychiatrists and the patient
organisation plead for a use of the chip in professional settings, while
some patients express their desire to use the lithium chip as a home-
care device. The developers are interested in creating revenue and are
exploring both of these options. When the lithium chip becomes available for use as a home-care medical device, the current responsibilities

and roles of patients will change. The patient will have to perform specific actions and decide on whether or not to act upon the results.
Also, the role and responsibilities of psychiatrists will change. Currently,
the laboratory sends the results of lithium checks to the psychiatrist.
In interviews, psychiatrists emphasised that one of their main responsibilities is to remain in control, meaning at least being aware of the
actions that patients are taking in relation to their medication. Some
patients said that they are looking forward to using the lithium chip
at home because it could reduce travel time, and, by having their own
chip, they can experiment with their medication to find the optimum
level. Other patients, however, were reluctant to use the chip and feared
not interpreting the results correctly. These patients were also worried
that the chip would replace face-to-face contact with psychiatrists. The
patient organisation is of the opinion that the lithium chip should only
be used in a professional setting. According to them, most patients are
not capable of interpreting the results correctly.

The insurance company articulated difficulties with regard to reimbursement. Currently, only care provided in certified institutions, such
as hospitals, gets reimbursed. Furthermore, the insurance company is in
doubt about whether to reimburse all home-care measurements or only
those that are prescribed by a medical professional.

Dramatic rehearsal in practice

In 2010, as part of the Dutch Societal Dialogue on Nanotechnology
(2009–2011) (CIEMD 2011), I organised a CTA+ workshop with the aim
of providing an opportunity for actors to explore destabilisations and
identify possible new roles, responsibilities, norms and values in relation
to the development of the lithium chip and POCT in healthcare. Based
on my diagnosis of current uncertainties and ambiguities, I selected and
invited lithium chip developers, insurance companies, patients, patient
organisations, psychiatrists, scientists, and lab-on-a-chip technology
and laboratory employees. In the actual workshop, representatives from
all these groups were present, except for laboratory employees. Through
snowball sampling, I came across an innovation consultant (a potential
participant that I had not identified based on my diagnosis) who lob-
bied insurance companies to invest in the development of the lithium
chip. The insurance company that was present at the workshop provided resources for the development of the lithium chip. At the same
time, it also had to make decisions about reimbursement.

Together with colleagues from the University of Twente,[6] I developed a techno-moral future scenario (based on the methodology of

'controlled speculation') of what might happen in the near future (time-line from 2011–2015),[7] with regard to changing roles, responsibilities and mandates of groups of actors involved (see Box 9.2). Prior to the workshop, the scenario and a short description of the set-up of the workshop were sent to the participants as a way for them to prepare. It was mentioned that the workshop would be held under the Chatham House Rules.[8]

Box 9.2 Summary of the narrative that was written in a techno-moral future scenario. The narrative revolves around the changing roles, responsibilities and values of 'good care' in relation to the development and embedding of the lithium chip

In 2011, an alliance between an insurance company, psychiatrists and a patient organisation for people with bipolar disorder advises to use the lithium chip only in a clinical setting so that psychiatrists can stay in control. This advice has been developed based on the results of a pilot study conducted by the Dutch National Centre of Expertise on Mental Health and Addiction. The Ministry of Welfare, Health and Sports follows this advice, and starts to develop guidelines for the embedding of the lithium chip in clinical settings. A group of patients that participated in the pilot study does not feel represented. For them, the lithium chip contributes to the quality of life (more freedom, less dependency) and they demand to have their own lithium chip at home. They set up a foundation, called 'Heft in Eigen Hand' and approach foreign lab-on-a-chip developers with a request to develop a lithium chip. In 2012, a company in the Czech Republic, licensed by Medimate, is willing to develop the chip and sell it on the Internet. Psychiatrists are watching these developments with great discontent. From their perspective, patients using the chip at home undermines the authority and knowledge of psychiatrists. Some psychiatrists cease to treat these patients because the responsibility and accountability issues for home-care use are not defined properly. In 2013, medication poisoning occurs caused by home-care use of the lithium chip. Who is responsible—the patient or the psychiatrist? The medical disciplinary tribunal judges that the psychiatrist is not to blame. When point-of-care devices are used at home, patients are the ones responsible. In the course of

2014, this verdict leads to societal upheaval. Politicians publicly question the benefits of free market dynamics in healthcare and the increasing demand on individuals to be proactive about their own (mental) health. In 2015, the Ministry of Welfare, Health and Sport wants to restore the authority of the medical professional. The professional, and not the patient, should have the final responsibility for home-care testing. However, this incentive by the Ministry is too late because the Medical disciplinary tribunal has judged otherwise.

Preparing and rehearsing a joint future: Interaction dynamics in the workshop

Participants explored, and reflected upon, the narrative sketched in the scenario. Issues discussed most extensively were: (i) Should the lithium chip be used at home or in a clinical setting?; (ii) If the lithium chip is going to be sold on the free market, how do we take the characteristics of bipolar disorder into account in developing a protocol for correct use?—participants agreed that owing to this particular group of users, the lithium chip cannot be assessed following the same criteria as have been set in the past for other point-of-care devices, such as the glucose meter or blood pressure meter; and (iii) How should the lithium chip be reimbursed?

Some participants, with the help of some moderation, tried to develop ways to address problematic issues and imagined, in interaction, the possible consequences (i.e. new roles, responsibilities, norms and values) of executing particular lines of action. For example, the lithium chip developer, insurance company, the patient organisation and the psychiatrist developed in conjunction, a possible scenario for how the lithium chip could be used at home. New roles and responsibilities were developed in relation to each other and in relation to the lithium chip. They articulated that for them, 'good care' implied that patients should be educated in how to use the chip and the psychiatrist-at-a-distance should know about the results of each measurement. Options to add telecommunication to the chip were explored, for example Bluetooth, in order to keep psychiatrists informed. A problematic issue that the insurance company identified in this respect was whether or not insurance companies should also be notified on the results of measurements. The representative of the insurance company articulated that having these data is a way to create personalised insurance packages, but he also wondered whether it would lead to a desirable society and healthcare. Prior to the workshop the psychiatrist was sceptical about the added value of the lithium chip in home settings. The only added value he saw was the use of the lithium chip to provide on-the-spot measurements in clinical settings. By hearing details about the functionalities of the lithium chip and by actively listening (listening to understand) to the experiences and desires of a patient (ED), the psychiatrist started to reflect on his role and responsibility as a psychiatrist. In interaction, the patient and the psychiatrist explored their possible new roles and responsibilities to deal with an emerging future situation of patients using the lithium chip at home. The patient (ED) articulated his desire to use the lithium chip at home because he wanted to experiment with his dosage in order to find the optimum. The psychiatrist responded by questioning what the added value is of experimenting at home instead of in a clinical setting. The patient answered that, for him, the main driver is to become less dependent on the psychiatrist. He (ED) continued, in more general terms:

> *00:52:19-9 ED*: Imagine if you don't have a good relationship with your psychiatrist, then...
> *00:52:20-3 I*: Then you should find another psychiatrist.
> *00:52:26-2 ED*: Finding another psychiatrist in Friesland is not that easy.
> *00:52:33-4 P*: Yes, that is a problem.

The following interaction sequence shows how the psychiatrist, after first inviting the patient to articulate his opinion further, starts to imagine how he can adapt his current role and responsibility to deal with new circumstances and challenges:

> *00:58:16-2 P*: Can you tell us more about how you perceive the relation between the client, the psychiatrist and the new device? How will the device change the relation between the client and the psychiatrist?
> *00:59:05-6 ED*: Well that you can perform your own measurements on a more frequent basis. You already have a life chart with results.[9] I mean, with more frequent measurements psychiatrists might be able to give a more detailed and better diagnosis.
> *01:00:00-8 P*: Well you are saying interesting things because I also have patients with whom I agreed that they can experiment with

their medication, in some situations. These patients might profit from using a lithium chip at home, in particular cases, for example when they are at risk. So not for all patients, but for a few patients, the lithium chip might be...

Later on in the workshop, the psychiatrist articulated his role and responsibility in relation to the new situation in more detail:

> ...it is, of course, a free market and clearly I cannot prevent patients to buy a lithium chip, but as a psychiatrist, I should have control and therefore I need to have face to face contact with my patients (...) However, when somebody wants to be treated by me in Groningen but is living in Beegrum, which is quite far away, then the lithium chip might be a solution to establish a relation with that person because he cannot come to Groningen every time...

The way in which the patient and psychiatrist (inter)acted, corresponds to Dewey's principle of (inter)action and participation in the light of the unknown: responsibilities, values and norms 'emerge' through inquiry into present challenges,and by daring to put existing moral routines 'at risk' in order to explore new ones. However, this kind of interaction to prepare for a possible future situation did not occur often during the workshop. In general, most interactions between the participants occurred in a transmission mode. Participants did not really question one another's opinions, values and dilemmas, but responded by expressing their own. Sometimes the moderator stimulated such questioning.

Effects of the workshop in the real world

To evaluate if, and how, participants learned from the CTA+ workshop and if, and how, they were prepared to act upon insights they gathered from the workshop, I conducted follow-up interviews one month after the workshop.[10]

One of the results was that, for all participants, the scenario stimulated learning and reflection on future societal practices that are in the making. The lithium chip developer and one of the patients indicated that the actual interactions in the workshop did not contribute much to increase their understanding of what is at stake. This may be related to the fact that these two actors, together with the chairs of the patient organisation and insurance company, already knew each other's positions and perspectives, being members of the advisory board set up by the developer of the lithium chip. Inspired by a CTA workshop organised in 2006 by van Merkerk (2007), the developer of the lithium chip decided to start an advisory board with potential users and an insurance company to discuss the development and embedding of the lithium chip. Referring to this advisory board, the lithium chip developer states:

> The workshop interactions did not really offer new insights. But I thought, well the scenario absolutely opened my eyes. For example the way you described how foreign companies, licensed by us, could sell the chip on Internet (...) that is something, I distributed the scenario to my colleagues and told them to be aware, this is what is happening and we should prepare ourselves.

Other participants also indicated that they took up insights from the workshop in their daily practices. The psychiatrist discussed insights from the workshop with his colleagues. The professor initiated actual action. In 2010, she was co-organiser of a large international conference on lab-on-a-chip-technology. She invited me and the moderator to organise a round-table discussion about the ethical and societal aspects of lab-on-a-chip technology because she anticipated that her colleagues would be interested in hearing about it.

Although the professor had not been active during the workshop, she mentioned that she did learn quite a bit from the workshop.

> The workshop was really fruitful for me. I work on lab-on-a-chip technology for more than 20 years now. But, to be honest, I never thought about the fact that people are actually going to work with lab-on-a-chip and that it can change healthcare practices.

This position of the professor is illustrative of the traditional position of scientists operating solely on the laboratory floor and not taking broader issues into account. But it is not a message of despair. The CTA workshop made a difference in making her enthusiastic and reflexive about her current role and responsibility. For example, she referred to how some participants found that too many promises are attached to the lithium chip (and to lab-on-a-chip in general). She relates this issue to her own task and responsibility as a scientist:

> Because of the workshop, I started thinking about our role as scientists. I think we have to leave our lab on a more regular basis

to communicate and inform people about the possibilities. I think the expectations about lab-on-a-chip are too high now. We probably also have to interact with end-users in an earlier stage so we might develop a different technology.

Discussion and reflection

Participants appreciated the workshop, which implies that the intervention (organising such a workshop) was considered legitimate. Was it also effective as a place where actors could explore and anticipate how their activities are entangled and which future roles, responsibilities, norms and values are emerging?

Learning with regard to the co-evolution between the development of point-of-care-testing and changing roles, responsibilities and mandates did take place. The professor and the psychiatrist articulated that this learning was an immediate effect of the actual interactions during the workshop. For all participants, learning was stimulated by the scenario, thus by external input. This is not the kind of dramatic rehearsal that Dewey envisaged. In his philosophy, learning takes places and understanding is generated through specific forms of interaction and participation.

Most participants were concerned primarily with sending their messages, rather than reacting to the opinions and experiences of others. They did not put their identities (values, norms, responsibilities) at risk during interactions in order to explore new ones. Instead, most participants continuously articulated their already established roles, values and responsibilities. When exploration and reflection on roles and responsibilities did occur, it was most often as a result of explicit moderation.

Dewey assumes in his work on dramatic rehearsal that people take up insights and test the applicability of new (or adapted) moral frameworks in real life. After the CTA+ intervention, only a few participants experimented in practice with possible new roles and responsibilities. This is not surprising considering the fact that the workshop was just one intervention. As soon as participants leave the workshop and go back to their own institutions they are confronted with constraints and exigencies, including the rules and repertoires, that function in their group or sector.

Does this mean that a dramatic rehearsal stimulated in a temporary, more or less protected, CTA+ workshop is not effective? Not necessarily, but, as argued by Rip and Shelley-Egan (2010), in our society, for changes in roles and responsibilities to occur, individual learning is not enough. There must also be openings at the collective level to take up insights and experiment with new roles, relations and responsibilities. The fact that some participants *did* discuss the insights of the workshop with their colleagues or initialised action shows that the constraints are not absolute.

How a particular newly emerging technology that is still in the R&D phase will shape societal practices is, to a large extent, still uncertain. However, current activities and choices of groups of actors involved enable and constrain which values, norms, roles and responsibilities might become dominant in the future, and which are silenced or remain under-developed. By moving about in the different 'worlds', observing and asking questions, an ethicist or social scientist can identify and articulate the first contours of how newly emerging technologies lead to destabilisations of existing societal practices. By organising a dedicated space for interaction, an ethicist or social scientist can foreground her/his diagnosis, and select and invite those who are involved to proactively explore and articulate what to value and how to act.

Designing and organising spaces where actors who normally operate quite independently from one another can meet and discuss the development and embedding of newly emerging technologies is not a prerogative of ethicists or social scientists. The case I presented showed that a technology developer can also organise a space. The prerogative of an ethicist or social scientist *is* that she/he can move around and therefore see things that are more difficult for actors operating in a particular domain, and focusing on realising their daily roles and responsibilities, to see properly. For Dewey, dramatic rehearsal functioned as a way for people to navigate more reflexively in a world in flux. For newly emerging technologies, dramatic rehearsal prepared by an ethicist or social scientist might do the same for groups of actors involved.

Notes

1. The defining characteristic of the midstream modulation approach is the presence of an 'embedded humanist' on the laboratory floor (Conley 2011; Schuurbiers and Fisher 2009) who tries to bring in societal considerations into a research laboratory. The 'embedded humanist' enters the laboratory as 'tabula rasa' (Conley 2011) and tries to generate understanding by engaging in the laboratory. The embedded humanist asks questions and probes the world of his or her dialogue partner (like Socrates did) with the aim of stimulating reflexivity and learning on (implicit) assumptions, values, norms

and frameworks, in this case in relation to decisions scientists make when conducting research.

2. For example, scientists and industrialists have a mandate to enact promises and expectations, and develop new science and technology because of their purported responsibility to work towards progress, knowledge production and economic prosperity. Other groups of actors, like government and funding agencies, have a task to compare and select between different options and develop regulation.

3. CTA, especially within the TA Nanoned programme (2004–2010), designed and organised dedicated spaces for interaction in which enactors and comparative selectors could explore mutual entanglements, force fields, ethical and societal issues, and where broader strategies could be developed by actors with the net effect of a better embedding of new technology in society.

4. ELSA is an abbreviation of ethical, legal, societal aspects.

5. Making the co-evolution visible and more reflexive served as the background to concrete aims in the TA Nanoned programme. My point is that the 'spotlight' (e.g. diagnosis, objectives and activities) of the TA Nanoned programme was primarily on the 'technology' sphere, i.e. on optimalisation of particular technology trajectories. The aim of CTA+ workshops is to inquire, deliberate and negotiate the desirability of possible future societal practices that are in the making owing to nanotechnology developments

6. I developed the scenario together with Marianne Boenink, Federica Lucivero and Arie Rip.

7. The workshop took place in 2010. At that time, 2011 and 2012 were, thus, part of the future. The lines of action sketched in the narrative are 'controlled speculation' and do not necessarily match actual developments that took place in 2011 and 2012.

8. This rule implies that 'when a meeting, or part thereof, is held under the Chatham House Rule, participants are free to use the information received, but neither the identity, nor the affiliation of the speaker(s), nor that of any participant, may be revealed' http://www.chathamhouse.org/about-us/chathamhouserule

9. A life chart is a systematic collection of past and current data on the course of illness and treatment recorded by the patient and/or clinician. Earlier in the workshop, the psychiatrist and the two patients articulated that if patients use the lithium chip at home, the results of each measurement can be linked to their individual life chart and/or with blue tooth, a psychiatrist-on-a-distance can be informed.

10. Based on criteria developed in TA Nanoned (Parandian, van Merkerk), I distinguished three forms of learning. Learning in the sense of (i) knowledge and/or better understanding of the technological device and its promises—did participants gain (more) insight into how the application works and for which problems does it propose a solution?; (ii) knowledge and/or better understanding of societal practices that are unfolding as a consequence of the development of the new technology—do participants reflect on own positions, responsibilities and mandates in relation to new societal practices?; (iii) knowledge and/or better understanding of assessment frames,

and responsibilities and mandates of the other groups of actors who are involved.

References

Boenink, M., Swierstra, T., and Stemerding, D. (2010) 'Anticipating the Interaction Between Technology and Morality: A Scenario Study of Experimenting with Humans in Bionanotechnology', *Studies in Ethics, Law and Technology*, 4(2): article 4; DOI:10.2202/1941-6008.1098.

CIEMDN (2011) *Responsibly Onwards with Nanotechnology. Findings March 2009–January 2011* (The Hague: CIEMDN).

Conley, S. N. (2011) 'Engagement Agents in the Making: On the Front Lines of Socio-technical Integration. Commentary on: "Constructing Productive Engagement: Pre-engagement Tools for Emerging Technologies"', *Science and Engineering Ethics*, 17(4): 715–21.

Deuten, J. J., Rip, A., and Jelsma, J. (1997) 'Societal Embedding and Product Creation Management', *Technology Analysis & Strategic Management*, 9(2): 131–48.

Dewey, J. (1920) *Reconstruction in Philosophy* (15th ed.) (Boston: Beacon Press).

Dewey, J. (1938) *Logic: The Theory of Inquiry* (New York: Holt, Rinehart and Winston).

Dewey, J. (1957) *Human Nature and Conduct: An Introduction to Social Psychology* (New York: The Modern Library).

Dijstelbloem, H. (2007) *De democratie anders: Politieke vernieuwing volgens dewey en latour* (Enschede: Iskamp).

Fesmire, S. (2003) *John Dewey and Moral Imagination: Pragmatism in Ethics* (Bloomington: Indiana University Press).

Garud, R. and Ahlstrom, D. (1997) 'Technology Assessment: A Socio-Cognitive Perspective', *Journal of Engineering and Technology Management*, 14(1): 25–48.

Hildebrand, D. L. (2008) *Dewey: A Beginner's Guide* (Oxford: Oneworld Publications).

Parandian, A. (2012) *Constructive TA of Newly Emerging Technologies. Stimulating Learning by Anticipation Through Bridging Events* (Delft: Technical University of Delft).

Rip, A. and Shelley-Egan, C. (2010) 'Positions and responsibilities in the "real" world of nanotechnology', in von Schomberg, R. and Davies, S. R. (eds) *Understanding Public Debate on Nanotechnologies, Options for Framing Public Policy* (Luxembourg: Publications Office of the European Union).

Rip, A. and Talma, A. S. (1998) 'Antagonistic Patterns and New Technologies', in Disco, C. and van der Meulen, B. (eds) *Getting New Technologies Together* (pp. 299–323) (Berlin: Walter de Gruyter).

Rip, A., Misa, T. J., Schot, J. (1995) 'Managing Technology in Society: The Approach of Constructive Technology Assessment' (London/New York: Pinter).

Robinson, D. K. R. (2010) *Constructive Technology Assessment of Newly Emerging Nanotechnologies* (Enschede: Iskamp Drukkers BV).

Schuurbiers, D. and Fisher, E. (2009) 'Lab-scale Intervention', *Embo Reports. Science & Society Series on Convergence Research*, 10(5): 424–7.

Sorensen, K. H. and Levold, N. (1992) 'Tacit Networks, Heterogeneous Engineers, and Embodied Technology', *Science, Technology, & Human Values*, 17(1): 13–35.

Swierstra, T. and Waelbers, K. (2012) 'Designing a Good Life: A Matrix for the Technological Mediation of Morality', *Engineering Ethics*, 18(1): 157–72.

Te Kulve, H. (2011) *Anticipatory Interventions and the Co-evolution of Nanotechnology and Society* (Enschede: Iskamp Drukkers, BV).

Van Merkerk, R. (2007) *Intervening in Emerging Nanotechnologies, a CTA of Lab-on-a-chip Technology* (Utrecht: Utrecht University).

Welchmann, J. (1995) *Dewey's Ethical Thought* (Ithaca: Cornell University Press).

Part III
Critical Perspectives

10
Pervasive Normativity and Emerging Technologies

Arie Rip

Introduction

Normativity is everywhere. It is taken up in ethics, but also in law and political theory, and more implicitly in economics and sociology. And there is the de facto normativity of master narratives and imaginaries,[1] like the modernist narrative of progress—in particular progress through science. There is also, underlying many normative issues, the fundamental challenge (*die ärgerliche Tatsache*) of social order, as a value in its own right, and thus to be conserved, and/or as a constraint that needs to be opened up. This essential ambivalence of social order feeds into the discourse about innovation, which can be embraced as wonderful, or criticised as deviant and possibly dangerous (cf. Godin 2010). Issues thrown up by emerging technologies partake in this fundamental challenge, and are thus broader (and deeper) than questions of risk and other immediate effects on society.

This is the first level of pervasiveness of normativity. One implication is that the achievements of ethics, 'the rich and diverse work of the past 2,500 years of moral philosophy' (Baggini and Fosl 2007: p. xvi), while considerable and to be built upon, cover only part of the articulations of normativity, in spite of the claim of ethicists that normativity is their special domain. I will discuss other articulations of normativity as in law, governance and sociology, and also consider 'sites' where pervasive normativity is visible. My own experience with constructive technology assessment (CTA) will be one such 'site'. This will also allow me to comment occasionally on what is sometimes called 'ethics on the laboratory floor' (see the introduction to this book), where normative reflection is stimulated by an ethicist or, more generally, a humanist (as Fisher phrases it in his STIR project; see Fisher and Rip in press).

On a second level, the question is what this normativity is supposed to be. I use the term to move away from normative as a qualifier, which creates a distinction with non-normative, to normativity as a basic feature, an anthropological category, which is the condition for the possibility of specific normativities. A similar move is made when 'sociality' is introduced over and above its dictionary definition of the tendency to form social groups and communities: there must be a basic anthropological category of sociality for this to occur at all.[2] Normativity is part of practices, not just in the sense that practices have tacit rules that can be articulated and reflected upon in terms of their status and justification, but because normativity is an integral element of practices and of social life generally. It is the condition of possibility to have rules and values that carry weight at all. In that sense, normativity is pervasive, as a basic category, as well as a competence. It is the soil on which the specialised normativities taken up professionally in ethics and law, and political theory can grow.[3] It is also the precondition for CTA philosophy (Rip and Robinson in press) and the possibility of reflective equilibrium (Daniels 1979; Rawls 1971; Thagard 1988).

For emerging technologies with their indeterminate future, there is the challenge of articulating appropriate values and rules that will carry weight. This happens through the articulation of promises and visions about new technosciences, like genomics and nanotechnology, and how these are responded to, sometimes querying them or raising concerns, and the further institutionalisation of the considerations and rules that emerge.[4] The content of the promise refers to desirable futures, but there are other issues as well, such as the question of plausibility of the promise given the essential uncertainty about their being realised, and the fact that their realisation does not just depend on the actor voicing the promise, but on many others, later in the chain of development and embedding in society. Whether the actor voicing the promise will be praised or blamed depends on the outcome of this co-production process, and thus represents a case of what is called 'moral luck' in the ethics tradition (Baggini and Fosl 2007: pp. 222–4).

There is a further consideration about how pervasive normativity works out in practice: the importance of second-order normative articulations. An example is how promises about an emerging technology are often inflated in order to get a hearing. Such exaggerated promises are like confidence tricks and can be condemned as bordering on the fraudulent. But then there is the argument that because of how science and innovation are organised in our societies, scientists are almost forced to exaggerate the promise of their envisaged work in order to compete for funding and other resources. The first-order normative articulation is about scientists exaggerating promises (and about science sponsors listening, and accepting them, to justify investing in the new possibilities). One can also inquire into the patterns and the mutual dependencies involved, like a game actors play in the real world (Scharpf 1997), and ask what the justification (or lack of justification) of this pattern might be. That would constitute a second-order normative articulation. In later sections I will offer further examples, which will reinforce my claim that second-order normative articulation is important.

In the next section, I will develop the point of pervasive normativity by considering relevant disciplines and the issues that are taken up. Emphasising pervasiveness of normativity as an anthropological category, so being everywhere and having various articulations, also implies an invitation to look for normativity in all sorts of places, and with fresh eyes. I will discuss two sites where normativity occurs. Firstly, the open-ended character of emerging technologies and the promises, and perhaps disappointments, that go with it. And, secondly, the way our CTA activities on newly emerging nanotechnology raise normative issues about these activities and the role of CTA agents. This will allow me to reflect about how to address pervasive normativity in relation to emerging technologies.

Pervasive normativity as articulated in relevant disciplines

Law, political theory and sociology all take up normativity, and address it in their disciplinary frames. But these frames are not given once and for all, and one sees further developments and some overlap, as in the recent interest in governance. I will briefly (and thus only partially) discuss such developments, and then comment on the ways ethics takes up pervasive normativity.

In law, some authors have emphasised how it relies on normativity as an anthropological category. Selznick (1992: p. 453) introduces the notion of 'implicit law', as the latent purpose, value premise or factual assumptions.[5] In contract law, for example, there are value premises about consent as the ground of obligation, facilitation of exchange, and good faith in negotiation and performance. Selznick then continues, and says 'In the quest for implicit law the most important move is the identification of values that are fundamental to a sphere of life or to the community as a whole. When such values are made explicit they can serve as authoritative guides to interpretation'. In common law (in the

UK and the USA), as well as in continental European tradition, there are openings to do so, but the articulations most often remain ad hoc.

Normativity in law is visible in the classical deontological approach and in another way—in the more recent positive approach to law as being authoritative by convention (because one can inquire into the status and justification of the conventions). There are now also the ideas of reflexive law (Teubner 1983),[6] and the related notion of responsive law (Nonet and Selznick 1978), which are important because they can address corporate social responsibility and the governance of the open-ended character of new technology. Teubner starts by distinguishing substantive law which specifies, for example through standard-setting or by outlining how (policy) objectives should be achieved, how people should behave from formal law, which stipulates, following the liberal tradition, procedures, but leaves actual value specification to the actors (e.g. negligence is stipulated as an occasion to call for compensation or redress, but without specifying what constitutes negligence). Reflexive law, as Teubner outlines it, does not specify what actors should do, but provokes reflection and auto-correction, still with some reference to the goals of the regulators.

From the point of view of reflexive law, the present interest in responsible research and innovation (cf. Fisher and Rip in press) constitutes a further step where the goals of the regulatory authority are no longer the criterion, but a good faith effort of the regulees is sufficient. In the same vein, the European Union Code of Conduct speaks of the importance of a 'culture of responsibility for us all'.[7] This indicates a move from law to governance, and the move is reinforced by the interest in horizontal interactions between old and new stakeholders (and publics). There are challenges for political theory here. For example, I have noted that the responsible research and innovation discourse and practice, when stabilised, will create a governance arrangement where stakeholders among themselves can, and will, decide about issues (about emerging technologies) that are in the public interest. This amounts to a return of neo-corporatism (which has never been fully absent).[8] Phrasing it this way, it is clear that there are links with political theory, but I will not pursue these. Instead, I will discuss governance—in general and in relation to emerging technologies.

The concept, as well as emerging practices of governance, includes law, political science and sociology. While often used as an analytical concept, there are always normative overtones, and these are explicit in the concept of 'good governance' as used by the World Bank when it sets conditions for supporting developing countries or when firms are evaluated as to their being 'good firms'. The reference to 'good' may lead to specifications of what is 'good' as in the deontological tradition, but also be taken as an occasion to articulate and debate what is to be considered as 'good'. There is a proactive, more or less top-down, route to do so, relying on parliamentary democracy and its extensions, and on the broader interactions in the public sphere. Another, complementary, route is to recognise how de facto governance occurs all the time and ask how such processes can be modulated. This is where sociology can contribute.

Thus, there are good reasons to use a broad definition of governance, which does not depend on a governing actor. 'In the broadest sense of the concept of governance, all structuring of action and interaction that has some authority and/or legitimacy counts as governance. (...) Governance arrangements may be designed to serve a purpose, but can also emerge and become forceful when institutionalised' (Rip 2010). The same move is visible in Voß et al. (2006: p. 8) when they argue that governance refers to 'the characteristic processes by which society defines and handles its problems. In this general sense, governance is about the self-steering of society'.[9]

The contribution of sociology plays out in two ways. Firstly, sociology offers diagnoses of overall developments, for example as when Ulrich Beck, in discussing the risk society, speaks of 'organised irresponsibility'—a diagnosis of our time indicating what has to change, somehow (Beck 1992).[10] Beck then turns to reflexive modernisation as the way forward (Beck et al. 1994). For my questions, Beck's early approach is important: just as one can speak of division of labour and analyse it sociologically (also as a question of political economy), there is a division of moral labour. One example is how scientists continue to argue that they are responsible for doing good science, while its effect on society has to be taken care of by others.[11] This stance is under pressure now (cf. the policy move towards responsible research and innovation). The overall issue of divisions of moral labour and their intersections, in general and for science, technology and society, deserves the attention of sociologists, political scientists and ethicists.

Secondly, sociology can make a reflexive contribution (cf. also Brown 1989) by articulating the broader patterns and trends that impinge on concrete practices and may shape what actors decide to do.[12] This includes the question of how openings emerge and/or are created in given societal orders. The possibility of such a contribution has implications for the role of the sociologist (as well as ethicists and humanists). They can reinforce what is happening, or induce doubts, or better, a more distantiated, and in that sense enlightened, way of thinking. Or even contest existing arrangements, for good reasons.[13]

With the advantage of this discussion, I can come back to the role of ethics, in general and on the laboratory floor and other 'floors', as one articulation of pervasive normativity. There is a rich ethics literature, ranging from analytical to interpretive/hermeneutical approaches, with pragmatist ethics somewhere in between (or better, offering a third position). In practice, on the various 'floors', the rich tradition may be short-circuited because ethicists are expected to offer expert assessments and advice, and may succumb to the temptation.[14] Bio-ethics has become institutionalised that way, with its principles that can be applied in an almost mechanical way (cf. checklist ethics, as discussed in the introduction to the book).[15] Also on 'floors' other than the laboratory floor, for instance in decision-making about life sciences, there is a demand, often uncritically, for ethics as input in decision-making.[16] This reinforces an ethics deficit model (cf. note 3).

Still, it is important for ethics to have links with practices, here ranging from research work in the laboratory (and elsewhere) to strategy articulation of actors involved in emerging technologies, and to recognise that there are normativities involved that may actually contribute to ethics (as pragmatist ethics would emphasise). This requires ethicists to relinquish the notion that they have privileged access to normativity and can just expand on it. Of course, not all ethicists act in such a high-handed manner, but it is a bit of an occupational risk. Ethicist do have a privileged position because of their being versed in the ethics literature, just as other professionals or semi-professionals will have special competencies that can be drawn upon.

Pervasive normativity in emerging technologies and their promises

Newly emerging sciences and technologies live on promises, on the anticipation of a bright future thanks to the emerging technology, if only it would get enough support to materialise the promise. Conversely, big promises attract attention and may lead to articulation of concerns, already at an early stage. Thus, emerging technologies, their promises and the further articulation that occurs, partly through contestation, is a site where one can expect to encounter normativity at work.

Anticipations and the voicing of expectations are a general feature of social life. There are general issues, such as performativity of expectations, once they are around (up to self-fulfilling prophecies—or self-negating prophecies, for that matter). For emerging technologies, the realisation of their promise is in the hands of many others besides the original promisor. While she/he might eventually be praised for being far-sighted, the realisation is a matter of 'moral luck', as discussed in the ethics tradition. In the introduction I mentioned the moral issue of exaggerating promises, felt to be necessary to get a hearing amidst many other promises (which then will have to be inflated as well). In a sense, putting forward only plausible promises is a collective good as it avoids misguided investment of effort and eventual disappointment. But individual actors have little incentive to restrain themselves in the service of realising the collective good, so it will not be realised unless special measures are taken.

Thus, there is a pattern that leads to inflation of promises and subsequent disappointment, and the rules and mutual dependencies that constitute the pattern might be subject to (second-order) normative articulation. Actually, the effect is a recurrent hype-disappointment cycle. Such a cycle was originally identified and used by the Gartner Group for new information and communications technology products and product options (Fenn and Raskino 2008). It is now used more widely, also for nanotechnology (Rip 2006b). Figure 10.1, drawing on the visualisation made available through Gartner Group's website publications, shows the cycle of inflation, disappointment and, after the shake-out, the survival of more realistic options.

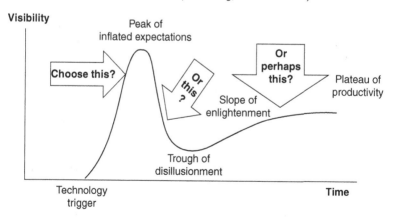

Figure 10.1 Questions raised by the hype cycle.

The cycle is presented as something that occurs all the time, i.e. as an empirical finding, but it is, first of all, a plausible storyline about how things tend to go. This then leads to the dialectics starting with patterns out there and, when recognised for what they are, actors profiting from them or modifying them, thus changing the pattern. If one 'knows' how things go, one can profit from that knowledge,[17] just as venture capitalists profit from their insight in the rise and fall of promising new firms. For example, get in while the hype is rising and get out before the disappointment sets in. Independently of whether there 'is' a recurrent pattern, the phenomena highlighted in the cycle occur and need to be addressed, not just as an opportunity or a concern, but also as a site where normativities are articulated.

One issue is choosing which expectation to pursue, for example by investing in it, or by putting it up as a priority, or by assessing it systematically. There are reasons to do so, but what may be neglected is that in a situation of limited resources, choosing one technological option means backgrounding other ones. Promoters of the backgrounded options may then raise their voice, and a process of articulation of arguments and values starts—as happens with alternative energy technologies (the choice for, and subsequent debate about, biofuels in the Netherlands and elsewhere would be an example). Such choices have to be addressed by technology assessment (TA) and also ethics: Which promises and which concerns need to be taken seriously, and considered more closely? Figure 10.1 shows the three types of locations on the cycle for choosing what to assess. From a public interest point of view, it would be ideal if one could already focus now on the options that will make it to the plateau of productivity, after the shake-out. Phrasing it this way, it is also clear that it is very uncertain what these options will be, in general, and because which options will survive the hype and disappointment phases will depend on the process just as much as on the features of the technological option. In concrete situations, there are all sorts of practical and political reasons for choosing one or another option, including the possibility to look at a technology on the upward slope of hype, and seeking to deflate the expectations.[18] For ethics, with its tradition of creating fictive situations to explore ethical issues, there is the temptation to focus on the grander promises, say about ambient intelligence and human enhancement. Alfred Nordmann and I have called this speculative ethics, and argued that focusing on it is a waste of scarce ethical resources when there are so many issues that should be considered and taken into account in choices and decisions about on-going developments (Nordmann and Rip 2009). The ensuing debate is still going on; it indicates that ethicists, when linking up with practices, cannot escape normativity.

There are the general problems of anticipation, where action is proposed in spite of the uncertainties and ambiguities that are involved in anticipation. In the public realm (but also inside organisations) there is the specific phenomenon of contention between inflation (to get a hearing) and 'deflation' (to reduce noise) of promises, as well as the phenomenon of early warnings (early signalling). There is content involved (what is the case, what is plausible), but also interests and values. The contention may have the form of a battle between entrenched positions, as can happen in controversies, for example about the use of genetically modified organisms, but will generally lead to articulation of positions and values, and some stabilisation in the sense that a robust repertoire of arguments and values will emerge that actors will draw on.[19]

Specific patterns of moral argumentation occur in the debates and deliberations on new sciences and technologies, for example nanotechnology (Swierstra and Rip 2007). They show how the new and promising is emphasised at first, but, when queried or criticised, the promisors revert to saying it's really nothing unusual (e.g. nature is genetically modifying all the time). The critics then come back with the slippery slope argument: maybe there is no problem now, but if we continue in this direction, who knows where we may end? So better stop now… Consumer thinking about new nanotechnology has been documented by Throne-Holst (2012): 'nanotechnology comes with risks; yes, but old technologies have risks as well; and nanotechnology may have advantages over old technologies'. These patterns are somewhat schematic, but they are clearly part of our cultural repertoire by now, and build on pervasive normativity, while articulating it further in concrete cases.

The key issue, which sets things in motion, is how to handle something new and the possibly exaggerated claims about it. This issue is not limited to emerging sciences and technologies, and ways to handle it have evolved. For advertisements, there are overall rules, and complaints about exaggerated claims can be lodged with a competent authority who will then rule on the matter. In the world of research on new drugs and therapies, there are now informal rules on how to publish preliminary findings, and how to avoid publicity that may raise possibly unfounded expectations with patients. Media of the more responsible kind have begun to report on new breakthroughs by including some words of caution. A clear example, showing the dialectics of promising, is a report in *Time* (December 3, 2007) on a breakthrough in stem-cell research by using adult cells rather than embryonic stem cells:

> Stem cells generated by this method are ideal not just because they are free of political and moral baggage. They can also be coaxed into becoming any type of tissue, and then be transplanted back into the donor with little risk of rejection.

After the thesis comes the anti-thesis:

> Still, these cells are far from ready for medical use. (...) Both Yamanaka and Thomson admit that we still know too little about how the process works to exploit the method's full potential.

A possible synthesis is visible in the rhetorical flourish with which the article ends:

> Nevertheless, their discovery has moved stem-cell research back to an embryonic state of its own – in which anything, it seems, is possible.

There are no informal rules for emerging technologies in general, but the issues have been visible in various practices. For example, around 1980, when biotechnology moved from open-ended promises about recombinant DNA to the discussion of first applications, the promoters of biotechnology followed two different strategies: emphasise the promises to mobilise resources, and thus increase the chance of realising at least some of the promises; or be modest so as not to run into disappointments which could lead funders and the industry to turn away from biotechnology altogether.[20] For nanotechnology, in the past decade, the USA tended to go for big promises, like a third industrial revolution and human enhancement, while in Europe the slogans were a little smaller and a little better. In both regions, there is now a new element in the repertoire, the discourse of responsible research and innovation (Fisher and Rip in press).

What we see happening at this site of pervasive normativity is increasing reflexivity about handling emerging technologies and their promises, and more explicit articulation of strategies, including normative aspects. These aspects are not always labelled as ethics, and ethicists are only beginning to pay attention to them. My point, then, is that it is more important that normativity is recognised and part of deliberation and action, than how it is labelled.

The site of CTA in action

The approach of CTA, with its emphasis on broadening technological development and its embedding in society by including more aspects and more actors is by now fairly well developed (Rip and Robinson in press). It has come in for critical comment, for example about its purported neglect of ethical issues (Lucivero 2012) or about its failure to substantially address democratisation, which is said to be (or should be) its goal (Genus 2006). We have emphasised that CTA is not about democratisation as such, but about 'a better technology in a better society' (Schot and Rip 1997), and about increasing reflexivity of co-evolution of science, technology and society [see Rip and Robinson (in press) for a discussion of goals and rationales of CTA]. Thus, there is discussion about the explicit and implicit normativity of CTA as an approach, as an effort to achieve certain goals. In this section, I reflect on CTA in action, as a site for reporting on normativity articulations as we encountered them, and how in handling them we had to articulate the normativity in our own position and roles.

CTA for emerging technologies offers special challenges.[21] The promise of new performance is often still a fiction, and assessment of future effects becomes social science fiction. For CTA of nanotechnology, where we now have a lot of experience (Parandian 2012; Robinson 2010; Te Kulve 2011), we have developed tools and approaches to turn the general discussion of promises into concrete articulation of endogenous futures, in sociotechnical scenarios informed by detailed study of the domain and by the insights from innovation studies and sociology of technology. Such sociotechnical scenarios then function in strategy articulation workshops as a support to different actors to articulate better strategies (of their own) for embedding emerging (nano)technology in society. In preparing such workshops and holding them, we interact with 'enactors' of the new technology and other stakeholders, we move about in the nano-world and encounter actor's normativities.[22]

An important de facto normativity is how promoters of nanotechnology (the 'enactors' of the promises) project scenarios of a brave new world thanks to nanotechnology, and evaluate what is happening in terms of acceptance of or resistance to these promises. Resistance is perceived as deriving from lack of understanding of nanotechnology and/or so-called 'emotional' responses to new technology.

The 'enactor' perspective can be an enlightened perspective, where the problems are not just attributed to the world out there, but also to own strategies, which must then be improved. There is now a general sense in the nanotechnology world that the impasse about genetically modified crops should be avoided ('This time we'll do it right from the very beginning').[23] A typical example of a somewhat enlightened perspective is visible in the presentation of Vicky Colvin, Director of the Center for Biological and Environmental Nanotechnology at Rice University, Texas, during the April 2003 US Congress hearings about the proposed nanotechnology bill. She used her own version of the hype-disappointment cycle to argue that Congress should help by requiring work on ethical, legal and social aspects (ELSA) of nanotechnology in parallel to the support of nanotechnology development:

> ... new developments in technology usually start out with strong public support, as the potential benefits to the economy, human health or quality of life are touted. At our center we call this the 'wow index'. (...) At present, nanotechnology has a very high wow index. For the past decade, nanotechnologists have basked in the glow of positive public opinion. We've wowed the public with our ability to manipulate matter at the atomic level and with grand visions of how we might use this ability. All this 'good news' has created a growing perception among business and government leaders that nanotechnology is a powerful platform for 21st century technologies. The good news has given nanotechnology a strong start with extraordinary levels of focused government funding, which is starting to reap tangible benefits to society.
>
> However, every new technology brings with it a set of societal and ethical concerns that can rapidly turn 'wow' into 'yuck'. [example of genetic manipulation of crops, public backlash crippling the industry despite the lack of sound scientific data about possible harm] The failure of the industry to produce and share information with public stakeholders left it ill-equipped to respond to GMO detractors. This industry went, in essence, from 'wow' to 'yuck' to 'bankrupt'. There is a powerful lesson here for nanotechnology.
>
> [This is the version from the hearings (Colvin 2003); a published version, with slight modifications, is available in Kulinowski 2004)].

She continued by saying, in essence: 'Please, Congress, force the nanotechnology enactors to do better'.[24]

The better behaviour, in Colvin's presentation, is defined as having ELSA studies accompany funding and stimulation of nanotechnology.[25]

Education of the public, and some public engagement, is the other commonly proposed better behaviour, even if there is no check as to whether this was actually missing for genetic modification in agriculture in the 1980s and 1990s. Clearly, the storyline about the so-called impasse she presents has become part of an accepted repertoire about what can happen to new technologies.

CTA workshops, and the whole approach of CTA, emphatically include the more or less enlightened enactors of nanotechnology. As they are the ones who can make a difference at an early stage of development, they constitute the primary target group, and CTA exercises have to cater to them without identifying with them. The construction of scenarios by developing 'endogenous futures' speaks to enactors because of their perspective (the scenarios, when done well, turn out to speak to other stakeholders as well, though).[26] We do introduce the 'shadow' of society: enactors know they have to take society into account, but do not really know how to do it or how to do it productively. Their first-round expectation about a CTA exercise is that it will offer them tools and approaches to do better.

But then, the idea is that enactors have to change their ways, at least to some extent, if better technology in a better society is to be achieved. While our CTA exercises put enactors in the centre (and could be criticised for that), we try to move them out of their enactor zone. Our attractive package has 'barbs' in it: in the scenarios by showing unexpected repercussions of actions of enactors and in the workshops by having other (and critical) stakeholders participating as equals. There will be mutuality in the 'probing of each other's realities' (Robinson 2010, after Garud and Ahlstrom 1997). The idea is that enactors cannot just continue in their own way. At least, if they do, they cannot say they do not know about alternatives.[27] In our evaluations of CTA exercises, we find that some enactors do move a bit, even if it is too much to expect them to change their ways.

Thus, the CTA agent has a changed perspective, working through 'soft intervention', as we have called it. This occurs not only through dedicated workshops and their preparation, but also through our insertion in the nano-world generally (Rip and Robinson in press). The CTA agent is a visitor, and has to be recognised as such, but as a knowledgeable visitor. Her/his visits call attention to the possibility and desirability of broadening technological development. There are various styles of doing this, demonstrating knowledgeability, but also including some playfulness as in telling stories about possible futures. Given the dominance of the narratives of progress and innovation, our CTA exercises adding some concrete question marks to these narratives, undermine the existing order a little bit, as a trickster does.

The trope of a trickster—or perhaps of the court jester who is accepted because thought not to be threatening—is important as indicating one mode of soft intervention, but the main value underlying the investment in CTA exercises is the value of reflexivity. This was made explicit in Schot and Rip (1997) when they concluded that the CTA agent is a reflexivity agent. How to justify this stance? It requires what I have called second-order normativity to address this question because it is about more than holding reflexivity as a personal value. In the large, it has to do with a diagnosis of how our social order evolves, and might become more reflexive. In the small, it is about how to interact with dominant enactors as part of making the co-evolution of science, technology and society more reflexive (Rip 2006a).

Reference might be made to the diagnosis of reflexive modernisation by Beck and others (Beck et al. 1994), but one should avoid Beck's eschatological claim about the coming of a second modernity (Beck 1992; Beck and Lau 2005). Reflexive modernisation is a broader trend, and Beck's idea of organised irresponsibility is important to understand in the presently fashionable discourse of responsible research and innovation: it attempts to overcome the organised irresponsibility that goes with emerging technologies.[28] New responsibilities will be articulated and may stabilise anyway. Thus, the call for responsible research and innovation opens up existing divisions of moral labour, but the closure can still go in different directions. One possible new division of moral labour might actually be one where reflection and articulation of normativity is delegated to ethicists, ELSA scholars and CTA agents (Rip 2009), and this is reinforced because it opens up new 'business' for these actors.[29] Thus, there is further issue, an issue of second-order normativity: inquiring into the value of this emerging new division of labour and what this implies for the responsibility of CTA agents (or ethicists, for that matter).

What I tried to show in this section is normativity at work, in the sense of perspectives of enactors with normative implications, and in how our encounters with enactors forced us to reflect on our soft interventions and the values involved. One conclusion is about showing oneself to be knowledgeable, so as to get a hearing and remain independent. Another conclusion is about our scenarios with the stories they tell (actually, we tell) about possible futures: these are small narratives that concretely question the big narratives about the importance of pushing science and technology.

Concluding reflections

The thinking offered in this chapter originated from our work with emerging technologies and our reflections about our work. This experience led me to consider normativity as an anthropological competence (not necessarily always leading to adequate performance), and take its pervasiveness as the starting point of the chapter. I indicated its articulation by professional scholars in ethics, law, political theory and sociology (and economics, but I did not discuss their special take on normativity). Then, I returned to emerging technologies, with their promises and with the attempts at soft intervention by CTA agents, and discussed them as sites where one sees normativity at work.

Such an exploration does not lead to formal conclusions. It does show the richness of pervasive normativity even while always fractured through institutions and roles and narratives. That is how normativity works out in the real world. Ethicists (i.e. mainstream ethicists) might not feel comfortable with this route, and put it down as embracing the social, at the expense of the normative. But the social is normative through and through, as I showed when discussing promising technologies and CTA. And it raises further important normative issues, in what I called second-order normativity (e.g. about divisions of moral labour), that have not been addressed by ethics. There is actually a third-order normativity: the importance to act and interact from a sense of open and rich normativity, and being willing to find it in unexpected places (up to tricksters playfully undermining social order).

Articulation of normativities in and around emerging technologies is occurring already, and becomes more explicit because of the involvement of ethicists, CTA agents and also because of the interactions of actors with different perspectives. Articulation here is not just reflection and discussion. Interactions and practices also become more articulated through action.[30] These actions are played out in multi-actor, multi-level processes full of uncertainties. Pragmatist ethics may be able to capture the normativities involved, but has to overcome its micro-level focus. This will allow a link with political theory, governance and (soft) law, and with agents creating visions, interacting and making satisficing decisions under uncertainty. While normativity is everywhere, it needs to be articulated to come into its own. In showing what happens at the sites of promising emerging technologies and CTA in action, I contributed to that articulation, without offering strong normative conclusions. This is in the spirit of pragmatist ethics, where normative positions co-evolve. The temptation to specify must be resisted because in interaction with the demands from technology actors it would lead to versions of checklist ethics, reducing the richness of pervasive normativity.

Acknowledgement

I am grateful to Tsjalling Swierstra for incisive, but helpful, comments on earlier versions of this chapter, and for inspiring discussion generally; and to the PhD students, old and new, who shared my encounters with the issues discussed in this chapter.

Notes

1. As a science and technology studies-oriented expert group report to the European Commission phrases it: 'All societies make use of characteristic, shared narratives that express wider imaginations about the world, how it functions, what is to be valued in it, and the place and agency of themselves and others in that world. These narratives are much more substantial than mere "stories"—they intersect dynamically with the material, institutional, economic, technical and cultural forms of society. They reflect prevailing institutional structures and reinforce collective aspirations. In worlds of policy practice, narratives, as we have observed throughout this report, tacitly define the horizons of possible and acceptable action, project and impose classifications, distinguish issues from non-issues, and actors from non-actors' (Felt et al. 2007: p. 73). The expert group report emphasises how existing imaginaries shape thinking and action. Charles Taylor (2002) shows how an imaginary of citizenship in the eighteenth and nineteenth centuries was creating (and subsequently stabilising) a public sphere in Western societies.
2. In offering this analogy, I am inspired by Knorr's recent work on object-centred sociality. 'Human beings may by nature be social animals, but forms of sociality are nonetheless changing (...). (...) new forms of binding self and other arise from the increasing role non-human objects play in a knowledge-based society and consumer culture and from changes in the nature of objects and the structure of the self' (Knorr 2007). Without emphasising it, she assumes sociality as a basic anthropological category ('binding self and other'), which can then take different forms.
3. In earlier work, for similar reasons, Clare Shelley-Egan and I have introduced the concept of ethicality. We wanted to argue against the ethics deficit model that is often the basis of attempts to enhance ethical reflexivity: the first step is then to offer ethics training to professional scientists and lay people. Our argument then was that ethicality is a competence of everybody (so an anthropological category), but not everybody is able to perform the competence equally well. So articulation of ethicality is in order, rather than introducing ethics into ethically empty vessels (Shelley-Egan 2011: Ch. 7). A similar move is made by Lindblom (1990) when he argues that 'inquiry' about our worlds, ourselves and our societies (in the sense of pragmatist

philosophers like John Dewey) is not limited to professional 'inquirers' like social scientists, although they do have a role to play because their scope of inquiry is larger, and the role of traditional views and cultural bias is less.

4. Thus, the success of emerging technologies depends on the emergence and stabilisation of such rules for their governance.

5. I am indebted to Bärbel Dorbeck-Jung for drawing my attention to the work of Philip Selznick (also in relation to the philosophy of CTA) and reinforcing my assessment of the importance of Teubner's work.

6. Teubner is inspired by inspired by Luhmann's sociological systems theory, but the point of reflexive law does not depend on Luhmann's theory.

7. 'A general culture of responsibility should be created in view of challenges and opportunities that may be raised in the future and that we cannot at present foresee' (European Commission 2008: p. 7).

8. Compared with earlier neo-corporatism, the indeterminacies of emerging technologies introduce new challenges because negotiations among stakeholder are not enough. There is also negotiation about future visions, which would lead to reflexive neo-corporatism. Actually, this may well be the best governance arrangement we can have for emerging technologies, as I claim in Randles et al. (2012).

9. They develop this further: governance is understood as the result of interaction of many actors who have their own particular problems, define goals and follow strategies to achieve them. Governance therefore also involves conflicting interests and struggle for dominance. From these interactions, however, certain patterns emerge, including national policy styles, regulatory arrangements, forms of organisational management and the structures of sectoral networks. These patterns display the specific ways in which social entities are governed. They comprise processes by which collective processes are defined and analysed, processes by which goals and assessments of solutions are formulated and processes in which action strategies are coordinated ... As such, governance takes place in coupled and overlapping arenas of interaction: in research and science, public discourse, companies, policy-making and other venues.

10. As Merkx (2008) has shown, organising of responsibilities occurs all the time, and definitely in response to new technological possibilities.

11. Such gerrymandering is captured in Ravetz's (1975) aphorism: Scientists claim credit for penicillin, while Society takes the blame for the Bomb.

12. Dorothy Smith (1988) characterised the specific role and craft of a sociologist, starting from the observation that phenomena are 'complex[es] of social relations beyond the scope of any one individual's experience' (p. 151). Although women and men 'are indeed the expert practitioners of their everyday worlds, the notion of the everyday world as problematic [rather than everyday world as another object of research] assumes that disclosure of the extralocal determinations of our experience does not lie within the scope of everyday practices. We can see only so much without specialised investigation, and the latter should be the sociologist's special business' (p. 161). This is not to claim a special, and exclusive, position for the sociologist, but to emphasise that a special craft is needed, which is not available in everyday practices. Anybody can try and develop this special

craft, and writers and travellers tend to do so to some extent. The sociologist *must* develop it, if she/he wants to do something worthwhile.

13. I have identified the role of tricksters (and court jesters, and 'coyote') in (Rip 2006a). The justification of such roles is a matter of second-order normativity, linked to a diagnosis of the quality of existing divisions of moral labour.

14. One example would be how the 'do no harm' principle is implemented by actors: 'Please, ethicist, tell me what is not allowed. I will take that into account, and then be free to do what I want otherwise'.

15. See, for example, Beauchamp and Childress (2001), then already in its fifth edition.

16. Felt et al. (2007: Ch. 4) draw attention to ways ethics, i.e. input from ethicists, is replacing law and politics in European governance, in bioethics and more broadly.

17. In fact, the Gartner Group has based part of its business model on the hype cycle by presenting and using it as a tool to offer strategic advice to companies. When you know, advised by the Gartner Group, where are you with your product(s) on the hype–disappointment cycle, you can define adequate strategies.

18. This is what I felt pressed to do in discussions and presentations for some domains of nanomedicine, like point-of-care diagnostics and nano-enabled targeted drug delivery (the 'magic bullet' promise). So substantial normativity of the CTA agent.

19. See, for further development of the dynamics of inflation/deflation struggles and the learning that occurs in controversies, McGinn (1979) and Rip (1986).

20. This is based on observation of internal discussions in the Programme Committee for the Dutch Innovation-Oriented (funding) Programme on Biotechnology, and on our collecting of documents and articles at the time. One finding was that scientists in disciplines far away from application, like molecular biology, would go for open-ended promises, while chemical technologists working on biotechnology would choose the modest strategy. For our overall evaluation of the Programme, see Rip and Nederhof (1986).

21. CTA addresses all sorts of technologies, not just emerging technologies; see, for example, the work on societal embedding of electric vehicles (Hoogma 2000; Schot and Rip 1997).

22. When I say 'we' I refer to myself and the PhD students working on CTA of nanotechnology in the TA programme of the Dutch research and development consortium NanoNed. Reflection on what we were doing, and on our implied and explicit normative stances, has been a shared concern, and some of it is visible in the PhD theses.

23. This type of phrasing can be seen in Roco and Bainbridge (2001) and Krupp and Holliday (2005).

24. The dialectics of recurrent patterns and actor strategies, as discussed in the section *Pervasive normativity in emerging technologies and their promises*, is clearly visible in the quote. The wow-to-yuck trajectory has to be painted as inevitable, so as to create a hearing for the message (the 'thesis'). But must then be reformulated as the result of mistakes, misguided behaviour, etc. (the 'antithesis') so that one can define better behaviour and go for it to achieve the desired goal, societal acceptance of nanotechnology (the 'synthesis').

25. She has also emphasised doing research on risks at an early stage (cf. Colvin 2005).
26. In principle, working with the sociotechnical scenarios runs into similar dialectics as in the Vicky Colvin quote, when offered as a support for actors to devise strategies to avoid futures they consider undesirable. The difference, however, is, firstly, that the scenarios are based on detailed empirical reconstruction of ongoing dynamics (rather than folk theories), and, secondly, that alternative futures are shown, so there is no recourse to a strong recurrent pattern (other than that the earlier pattern and the actor games involved might continue if nothing special happens or is done).
27. Parandian (2012) uses a loaded phrase to indicate this: they cannot use the argument of 'Wir haben es nicht gewusst' any more.
28. Compared with the earlier discussion of the responsibility of scientists, responsible now does not refer to actors who carry responsibilities, but to research and innovation.
29. This happened, to a certain extent, in NanoNed and its successor research and development program NanoNextNL.
30. For example, handling an artefact (which constrains and enables us) is articulating what it can do and does to us. So artefacts have morals, but this should not be reduced to the idea that they have been put into them by designers. Controversies lead to repertoire learning (at the collective level) which includes value articulation (Rip 1986).

References

Baggini, J. and Fosl P. S. (2007) *The Ethics Toolkit. A Compendium of Ethical Concepts and Methods* (Malden, MA/Oxford: Blackwell Publishing).

Beauchamp, T.L. and Childress, J. F. (2001) *Principles of Biomedical Ethics* (New York: Oxford University Press).

Beck, U. (1992) *Risk Society. Towards a New Modernity* (Cambridge: Polity Press).

Beck, U. and Lau, C. (2005) 'Second Modernity as a Research Agenda: Theoretical and Empirical Explorations in the "Meta-Change" of Modern Society', *British Journal of Sociology*, 99(4): 525–57.

Beck, U., Giddens, A., and Lash, S. (1994) *Reflexive Modernization* (Cambridge: Polity Press).

Brown, R. H. (1989) *A Poetic for Sociology. Towards a Logic of Discovery for the Human Sciences* (Chicago: University of Chicago Press).

Colvin, V. L. (2003) Testimony of Dr Vicki L. Colvin, Director Center for Biological and Environmental Nanotechnology (CBEN) and Associate Professor of Chemistry Rice University, Houston, Texas before the U.S. House of Representatives Committee on Science in regard to 'Nanotechnology Research and Development Act of 2003' (9 April 2003).

Colvin, V. L. (2005) 'Could Engineered Nanoparticles Affect our Environment?', in *Swiss Re Centre for Global Dialogue*, pp. 19–20.

Daniels, N. (1979) 'Wide Reflective Equilibrium and Theory Acceptance in Ethics', *Journal of Philosophy*, 76: 256–82.

European Commission (2008) 'Commission Recommendation on a Code of Conduct for Responsible Nanotechnologies Research', available at:http://ec.europa.

eu/nanotechnology/pdf/nanocode-rec_pe0894c_en.pdf (accessed 30 January 2013).

Felt, U., Wynne, B., et al (2007) 'Taking European Knowledge Society Seriously', report of the Expert Group on Science and Governance, to the Science, Economy and Society Directorate, Directorate-General for Research, European Commission (EUR 22700) (Brussels: European Communities).

Fenn, J. and Raskino, M. (2008) *Mastering the Hype Cycle: How to Choose the Right Innovation at the Right Time* (Boston, MA: Harvard Business Press).

Fisher, E. and Rip, A. (in press) 'Responsible Innovation: Multi-level Dynamics and Soft Intervention Practices', in Owen, R., Heintz, M. and Bessant, J. (eds) *Responsible Innovation* (pp. 165–83) (Chichester: Wiley).

Garud R. and Ahlstrom D. (1997) 'Technology Assessment: A Socio-cognitive Perspective', *Journal of Engineering and Technology Management*, 14(1997): 25–48.

Genus, A. (2006) 'Rethinking Constructive Technology Assessment as Democratic, Reflective, Discourse', *Technology Forecasting & Social Change*, 73: 13–26.

Godin, B. (2010) '$\kappa\alpha\iota\nu o\tau o\mu i\alpha$, Res Nova, Innouation, or, The De-Contestation of a Political and Contested Concept'. Project on the Intellectual History of Innovation Working Paper No. 9. Presented at the EASST Conference, Trento, 2–4 September 2010.

Hoogma, R. (2000) 'Exploiting Technological Niches: Strategies for Experimental Introduction of Electric Vehicles', PhD dissertation (Enschede: Twente University Press).

Knorr, K. (2007) 'Postsocial', in Ritzer, G. (ed.) *Blackwell Encyclopedia of Sociology* (Chichester: Wiley-Blackwell).

Krupp, F. and Holliday, C. (2005) 'Let's Get Nanotech Right', *Wall Street Journal*, 14 Jun.

Kulinowski, K. M. (2004) 'Nanotechnology: From 'Wow' to 'Yuck'?', *Bulletin of Science, Technology and Society*, 24(1): 13–20.

Lindblom, C. E. (1990) *Inquiry and Change. The Troubled Attempt to Understand and Shape Society* (New Haven: Yale University Press).

Lucivero, F. (2012) 'Too Good to be True? Appraising Expectations for Ethical Technology Assessment', PhD thesis (Enschede: University of Twente).

McGinn, R.E. (1979) 'In Defense of Intangibles: The Responsibility-Feasibility Dilemma in Modern Technological Innovation', *Science, Technology & Human Values*, Fall: 4–10.

Merkx, F. (2008) 'Organizing Responsibilities for Novelties in Medical Genetics', PhD thesis (Enschede: University of Twente).

Nonet, P. and Selznick, P. (1978) *Law and Society in Transition. Toward Responsive Law* (New York: Harper & Row).

Nordmann, A. and Rip, A. (2009) 'Mind the Gap Revisited', *Nature Nanotechnology* 4: 273–4.

Parandian, A. 'Constructive TA of Newly Emerging Technologies. Stimulating Learning by Anticipation Through Bridging Events', PhD thesis (Delft: Technical University of Delft).

Randles, S., Youtie, J., Guston, D., Harthtorn, B., Newfield, C., Shapira, P., et al. (2012) 'A Trans-Atlantic Conversation on Responsible Innovation & Responsible Governance', in Van Lente, H., Coenen, C., Fleischer, T., Konrad, K., Krabbenborg, L., Milburn, C., et al. (eds) *Little by Little: Expansions of Nanoscience and Emerging Technologies*, Proceedings of the third S.NET

Conference in Tempe, Arizona, November 2011. (pp. 169-179) (Dordrecht: AKA-Verlag/IOS Press).

Ravetz, J. (1975) '...et augebitur scientia', in Harré R. (ed.) *Problems of Scientific Revolution: Progress and Obstacles to Progress in the Sciences* (pp. 42–57) (Oxford: Clarendon Press).

Rawls, J. (1971) *A Theory of Justice* (Cambridge, MA): The Belknap Press of Harvard University Press).

Rip, A. (1986) 'Controversies as Informal Technology Assessment', *Knowledge*, 8(2): 349–71.

Rip, A. (2006a) 'A Co-evolutionary Approach to Reflexive Governance – and Its Ironies', in Voß, J.-P., Bauknecht, D., and Kemp, R. (eds) *Reflexive Governance for Sustainable Development* (pp. 82–100) (Chichester: Edward Elgar).

Rip, A. (2006b) 'Folk Theories of Nanotechnologists', *Science as Culture*, 15(4): 349–65.

Rip, A. (2009) 'Futures of ELSA', *EMBO Reports*, 10(7): 666–70.

Rip, A. (2010) 'De Facto Governance of Nanotechnologies', in Goodwin, M., Koops, B.-J., and Leenes, R. (eds) *Dimensions of Technology Regulation* (pp. 285–308) (Nijmegen: Wolf Legal Publishers).

Rip, A. and Nederhof, A.J. (1986) 'Between Dirigism and Laisser Faire: Effects of Implementing the Science Policy Priority for Biotechnology in the Netherlands', *Research Policy*, 15: 253–68.

Rip, A. and Robinson D.K.R. (in press) 'Constructive Technology Assessment and the Methodology of Insertion', in van de Poel, I., Doorn, N., Schuurbiers, D., and Gorman M. E. (eds) *Opening up the Laboratory: Approaches for Early Engagement with New Technologies* (Chichester: Wiley-Blackwell).

Robinson, D. (2010) 'Constructive Technology Assessment of Emerging Nanotechnologies. Experiments in Interactions', PhD thesis (Enschede: University of Twente).

Roco, M. and Bainbridge, W. S. (eds) (2001) *Societal Implications of Nanoscience and Nanotechnology* (Boston, MA: Kluwer Academic Publishers).

Scharpf, F.W. (1997) *Games Real People Play. Actor-Centred Institutionalism in Policy Research* (Boulder, CO: Westview Press).

Schot, J. and Rip, A. (1997) 'The Past and Future of Constructive Technology Assessment', *Technological Forecasting and Social Change*, 54: 251–68.

Selznick, P. (1992) *The Moral Commonwealth. Social Theory and the Promise of Community* (Berkeley, CA: University of California Press).

Shelley-Egan, C. (2011) 'Ethics in Practice: Responding to an Evolving Problematic Situation of Nanotechnology in Society', PhD thesis (Enschede: University of Twente).

Smith, D. E. (1988) *The Everyday World as Problematic. A Feminist Sociology* (Milton Keynes: Open University Press).

Swierstra, T. and Rip, A. (2007) 'Nano-ethics as NEST-ethics: Patterns of Moral Argumentation About New and Emerging Science and Technology', *NanoEthics*, 1: 3–20.

Taylor, C. (2002) 'Modern Social Imaginaries', *Public Culture*, 14(1): 91–124.

Te Kulve, H. (2011) 'Anticipatory Interventions in the Co-evolution of Nanotechnology and Society', PhD thesis (Enschede: University of Twente).

Teubner, G. (1983) 'Substantive and Reflexive Elements in Modern Law', *Law & Society Review*, 17: 239–85.

Thagard, P. (ed.) (1988) 'From the Descriptive to the Normative', in *Computational Philosophy of Science*, (pp. 113–37) (Cambridge, MA: MIT Press).

Throne-Holst, H. (2012) 'Consumers, Nanotechnology and Responsibilities. Operationalizing the Risk Society', PhD thesis (Enschede: University of Twente).

Voß, J.-P., Bauknecht, D., and Kemp, R. (eds) (2006) *Reflexive Governance for Sustainable Development* (Cheltenham: Edward Elgar).

11
Underdetermination and Overconfidence: Constructivism, Design Thinking and the ~~Ethics~~ Politics of Research

Alfred Nordmann

Introduction

'Ethics on the laboratory floor' is, at the same time, the triumph and *reductio ad absurdum* of constructivist accounts of scientific practice with their implied decisionism, if not voluntarism. It is the triumph of constructivism, as its insights are no longer thought to undermine science and its quest for reality or truth. Having exposed the myth that science follows for the most part, a collectively binding logic of research as envisaged by Karl Popper, as well as Thomas Kuhn, we now fully appreciate that scientists make choices. And these choices may carry the signature of prejudice or ideology, private interest or aesthetic preference, ethics or politics. The 'science wars' that revolved around the apparent abandonment of an image of science that speaks truth to power (Sokal 1996) have given way to a 'love fest' that celebrates the openness of science, for example to ethical consideration (cf. Nordmann 2007; Nowotny et al. 2003). But here, the proximity of triumph and *reductio ad absurdum* comes in. On the one hand, we celebrate the possibility of 'midstream modulation' of scientific practice where social scientists and philosophers induce reflectiveness about the relative environmental merits of using this material or that in a laboratory experiment (Fisher and Mahajan 2006). On the other hand, we thereby perform a vanishing trick which is very welcome to science policymakers: by fostering the illusion that our scientific and technological future is an aggregate of decisions on the laboratory floor, one arrives at a policy landscape where responsible innovation means as much as making sure that researchers are prudent and that citizens are well prepared for what it is to come in the future. In the meantime, and in the middle between these remote poles, there are no hard choices to be made about the directions of research, about funding priorities. On this account, the difficult choices take place where ordinary political debate does not reach, namely among experts on the laboratory floor or in the future when people will be confronted with new technologies. As for the here and now, public debate about emerging technologies is generally welcome, but need not address itself to any concrete policy choices. Accordingly, the role of policy is merely to create an environment that is generally conducive to innovation, that cultivates ethical sensitivities in the laboratory, and that fosters public preparedness for emerging technologies.

This is a caricature, to be sure, of current discourse about scientific and technological development. But it is telling enough that such a caricature can appear plausible (Davies et al. 2009; Nordmann and Schwarz 2010). And this is even more striking as one can so easily expose the proximity of triumph and absurdity in the fixation on decision-making within the day-to-day running of the laboratory (Winner 1993): as important an insight as it was that the decisions of researchers are underdetermined by logic and evidence, and thus open to societal considerations, it is, nevertheless, an untenable suggestion that these choices are of considerable consequence for technological development. The choice of material or method can be very important, no doubt, but doesn't quite measure up to the decision as to whether or not mortality from cancer should have higher research priority than chronic pain, or the question of balancing biofuels against agricultural crops, or the choice of mitigation and adaptation strategies in response to global warming. It is at the latter scale that the 'real decisions' about scientific and technological development are made, and this scale is incommensurate to that of the laboratory floor, if only because decisions about the direction of research set the frame in which research trajectories unfold, while decisions on the laboratory floor exploit the remaining degrees of freedom within the framework.

What follows from this critique of contemporary science policy discourse and its focus on scientific conduct? (cf. EC Code of Conduct 2008) One should not conclude, of course, that there is no place for 'ethics on the laboratory floor'. Reflexivity is rightly considered a virtue in almost any situation. Though it would be detrimental for public debate to abdicate its role in setting a research agenda, it is, nevertheless, only for the better if ethical awareness becomes integrated into the research process. As I hope to show, however, such ethical awareness should, almost by necessity, blossom into something more political: one of the first signs of their ethical awareness should be that researchers refuse the notion that their ethical awareness is particularly important. They would thereby also reject the presupposition that the solutions to our global and societal problems will emanate from science and technology (e.g. Vogt 2010). In other words, through their refusal they would reconnect their deliberations in the laboratory with political processes and cultural concerns.[1]

As long as we imagine the laboratory as a retreat for researchers and developers, the incommensurability of concerns will prove counterproductive. One way of breaking out of the laboratory, narrowly conceived, is to reconnect ethics and politics and to politicise research, that is make it subject to public contestation from any number of perspectives. Another way of absorbing laboratory activities within a larger process is to conceive scientific and technological research as part of a design process that begins with the definition of problems in society at large and that extends all the way to the users who settle upon the final shape of a technology (e.g. Chapter 4). From this point of view, questions of ethics and politics do not so much latch on to scientific or technological research and development, but organise it. They do not just inhabit the cracks and crevices where research and development are underdetermined by logic and evidence. If science and technology are part of a design process, questions of ethics and politics stand at the beginning with the question 'What kind of world do we want?', and accompany this process all the way up to the point where we deal with the ways in which the new technologies condition or mediate human interactions. But if this path is chosen and the larger view taken, the integration of ethics on the laboratory floor is preceded by another, more fundamental, philosophical task, namely the task to convince researchers that they have been recruited into a design process. Once they understand the changing character of research and once they understand their own place in the design process, they cannot help but appreciate the political, i.e. publically contestable character of the question 'What kind of world do we want?' (cf. Chapter 1).

So far, so good, and a lot more promising than what constructivist interpretations of science can offer—but to speak of research in a design mode (A. Johnson, personal communication 2009) also requires careful qualification. Where constructivist accounts overestimate the reach of the many minor decisions researchers make in the laboratory, this technological account of research tends to exaggerate human powers of shaping and controlling the world, and it avoids this exaggeration only at the expense of diffusing human agency to the point where the notion of 'design' becomes meaningless. In order to fully appreciate this dilemma, it is well worth engaging with Peter-Paul Verbeek's contributions on the topic, especially Chapter 4.

Two meanings of design

Verbeek's analysis begins with a view of technology as conditioning human sociability. Whether one argues with Kant or Foucault that technological infrastructure is a limit that makes certain kinds of experience possible, whether one argues with Winner and Wittgenstein that technologies are forms of life, or whether one argues with Latour or others that our conduct is, to a considerable extent, scripted by technologies, one arrives at the same conclusion: technological innovation alters the way we live and therefore involves conceptions of the good life at all times. From here, one quickly arrives at the notion that design is experimental ethics in that it engages competing conceptions of the good life. This is a compelling account, culminating in Verbeek's version of ethics on the laboratory floor: 'Responsible design requires an anticipation, assessment and explicit design of the mediations that the technology will introduce in society' (Chapter 4). Verbeek's account, in general, and this statement, in particular, invite two different readings, however, and each of these poses problems of its own.

On the first reading, 'design' refers to a kind of intervention such that collectively or individually, certain people induce changes that have far-reaching and wide-ranging effects: questions of ethics and responsibility arise in regard to these effects, presupposing the ability to predict and, ideally, control them. It is technological hubris, however, to assume that human designs are more than mere attempts made in good faith to alter reality, and to assume that they actually involve specific powers to produce at will, the intended, as well as the attending effects, of our interventions. It holds for even the simplest of actions that arise from the intent to bring something about and perhaps to change something for the better, and that it is executed despite the irreducible uncertainty as to whether our actions actually have the intended effect. On this first reading, however, this uncertainty of design action is suppressed, and a plan for the future is accompanied by anticipations of the future and further plans to control the anticipated effects, etc. Inversely, on this account we live in a world that is not just contingently or haphazardly dependent on human activity, but that is of our own making in the strong sense of being a world of our own design. Following Ihde (2008), Aslan Kiran refers to this view as the designer fallacy which mistakes the plasticity of reality and its openness to human tinkering with the idea that reality as a whole can be brought to agree with human wishing and willing (Kiran 2012). Undoubtedly, if mere wishing, planning and designing made it so, the ethics of wishing in the right way and for the right things would be paramount (Nordmann et al. 2011). But as this view rests on technological hubris and the designer fallacy, ethics is required not primarily for purposes of anticipation and control, but as a corrective that moderates hubris and conveys a humbling sense of contingency and limits of technical control.

The second reading is suggested by Ihde and Kiran's critique of the designer fallacy, and especially of its assumption that designers confront the world as if they weren't already deeply implicated in it through technologies that condition their perspective, that inform and structure their actions, and that will be modified through these actions. Instead of assuming an interventionist point of view that attributes responsibility for unidirectional actions and their resulting effects, one can overcome the designer fallacy by appreciating the multidirectional ubiquitous mediations of human collectives who constantly create and recreate a world that constantly creates and recreates them. This view is descriptively adequate, avoids technological hubris, does justice to the notion of the co-construction of technology and society, but undermines our usual notion of design in that it does not allow for the coordination of present actions to future states. In particular, Verbeek's idea that responsible design involves 'explicit design of the mediations' becomes intractable as those mediations resist objectification as targets of design that are under the control of a designer. Instead of signifying a controlled manner of shaping reality or giving form to it, 'design' now refers to an inextricable totality or pervasive medium of mediations in which we are called upon to conform to and thereby modulate our technological condition (Gamm 2000). In this rather unusual, perhaps counterintuitive, sense of the term, 'design' emerges as a proper translation of Heidegger's *Gestell*—technology challenges us, has designs on us, even as we go about the business of transforming the whole world into a resource by challenging it to deliver what, on scientific grounds, we demand that it yields (Heidegger 1993). As opposed to the familiar usage of the word, 'design' would now refer to a common fate that distributes human agency over a boundless variety of mediations.

Is there a middle ground between these two readings and these two meanings of 'design', one that would allow us to situate research ethics within the large-scale design projects of modern science and technology, or, rather, technoscience (Nordmann, et al. 2011)—avoiding the Scylla of technological hubris of technology assessment, namely the belief that we can actually and responsibly create the social world by getting the design process right, and avoiding the Charybdis of a widely distributed and generalised responsibility by everyone for everything that issues from all-encompassing technological mediations? In order to uncover this middle ground, we first need to discover a subject that has agency and bears responsibility and makes specific contributions to the design process. Second, we need to consider the objects of design, whether, how and by whom they are determined. Third, we need to question the designer fallacy, not by abandoning altogether the subject–object separation in the design process, but by exploring whether and when objects qua objects need to be viewed as being under the partial or complete control of some subject. Fourth, we need to justify this subject–object separation, if only as a necessary fiction. Only then can we return in a fifth and final step to Verbeek's most important insight, namely the idea that responsible design involves a conception of the 'morality of things' (Verbeek 2011). But, as we shall see, this understanding of responsible design reaches far beyond the laboratory floor. To pursue all five steps of this programme could take a long time and fill many pages. But a short sketch might do the trick for now.

Responsible innovation beyond the laboratory

The difficulty of identifying the designers and the impossibility of confining them to the laboratory floor comes out in Verbeek's contribution—and it resonates with the earlier critique of the constructivist approach: once you are in the laboratory, seeking to work out the proper diagnostic technique for a disease, to construct a better catalyst for this process or that, to extract biofuels from some plant or another, most choices have already been made, and the remaining scope of choice will be constrained by the evolved competence of the laboratory, by access to materials and techniques, by design specifications—which is not to deny that choices are made or that these can benefit from ethical reflexivity. However, an expanded conception of responsibility for the world and the lives that are constituted through design processes by necessity involves an expanded conception of the actors who contribute to the design process—it would include science policy-makers, market researchers, advocacy groups; it would include developers not only of devices but also of packaging and advertising; it would include early adopters and other buyers who establish the patterns of use that settle the definition of the designed artefact. To say that all of these are responsible for the mediations that result from their modification of the already-given technological infrastructure is not saying much, but does not make it impossible to identify particular duties or obligations toward which specific actors can be held accountable [cf. von Schomberg's proposal to establish relations of co-responsibility or responsiveness (2010)]. In the case of researchers this includes, famously, the question of whether they should even accept a design brief or not—a question that holds them to the high standard of opting out of their profession, if necessary. If the demand for such moral heroism should be reserved for extreme cases, one might construct the more general requirement that scientists should be good citizens, that is adopt a political understanding of the very design projects that they are contributing to.

On this construction of the many contributors to the design process, there is an object to their concerted efforts. It is not necessary, however, to conceive of this object as a persistent, homogeneous object that retains its definition throughout the design process. In her contribution to this volume, Bernadette Bensaude-Vincent provides an account of objects of technoscientific research that lack the precise determination of engineering artefacts or consumer products (Chapter 1). Instead, these objects are interesting for their potential and thus for their openness to receive more concrete determinations. Accordingly, if responsible researchers query the political brief that informs their work in the design process, they will often find that this brief does not require them to develop a specific product or a device, but rather to develop general capacities for innovation. A political understanding of their research project might then prompt 'good citizen' researchers to expose or render amenable for public debate the contestable cultural expectations that underwrite their design projects. For example, these projects may well aim for achievements of dematerialisation, of uncovering potentials, of enhancing material nature, or of transgressing presumed limits of resources or of space. A political critique of this design process would question the operative notions of innovation, sustainability, ecological and economic win–win situations, contrasting them to conventional wisdom and now possibly obsolete notions of conservation, and of a limited world that allows only for zero-sum games (Schwarz and Nordmann 2010). These 'good citizen' researchers would therefore amplify and publicise the ambivalence they experience, even in their laboratories: their design projects are underwritten by the expectation that science and technology can offer the necessary solutions to the problems of an otherwise unsustainable world—and though most researchers benefit from these expectations, they are simultaneously overtaxed by them (Vogt 2010).

On this kind of account, one can escape the designer fallacy, acknowledge the multiplicity of mediations, and, at the same time, retain a provisional separation of researchers as designers and their laboratory objects of interest. The relation between researchers and their object is not one of a controlled design process where plans and intentions become materially realised; it consists of the acquisition of capabilities of control that are conferred by the objects to the researchers and vice versa [cf. Latour's suggestion that researchers and their objects both acquire competences (e.g. Latour 1990)]. Though the design process is all about control, it must countenance limits of knowledge and control regarding desired and attending effects, intended and unintended consequences (see, e.g., Böschen and Wehling 2004).

So far, this has been an attempt to empirically carve out the middle ground between designer fallacy on the one hand, and the distribution of agency and responsibility over the boundlessness of mediations on the other. It views the design process as being unsurveyably comprehensive, proceeding in a manner that can hardly be considered deliberate. And it isolates from this process moments of intentional intervention that influence, but do not fully control the development of capabilities, artefacts or devices that will eventually condition human interactions, also those of the buyers or consumers who conclude the design process by settling the patterns of use. This empirical approach may be considered fine-grained, but does not dissolve or eliminate the conceptual tension between a world that is subject to design and a world of multiple mediations. This tension remains alive and is at work, for example in Kiran's attempt to salvage a notion of proactive design against Ihde's charge of technological hubris, even though '[p]roactive design implies constraining the ways a technology is handled in order to limit its undesired soft impacts' (Kiran 2012, p. 188). On the one hand, 'proactivity' refers to a heightened capability to predict, anticipate and intervene; it thus expresses the technological hubris of the designer fallacy. On the other hand, it is meant by Kiran to serve as a corrective to technological hubris and the designer fallacy, expressing something much weaker than control.

The notion of responsible proactivity, as well as the proposed empirical way of carving out the place of researchers in the design process needs to be complemented by a justificatory argument that may be required to extract subject and object from the maze of mediations, and to establish their antagonism in a design process. Such an argument could build on the Kantian idea of the limits that simultaneously constrain and constitute a world of experience and human interaction. In Verbeek's terms, the limit in question comes with a state of technology that gives rise to a world of mediations such that we can recognise this state of technology as constitutive of the social and natural world. However, the accurate description of this limit, that is of our entanglements in the social and natural worlds, implies that any particular human action makes hardly any difference at all in the bigger scheme of things. To ensure the very possibility of politics, however, one needs to produce an image of agency that allows us to see the world as being of our design—it is a precondition of politics or the necessary conceit that choosing to act this way or that can make a difference. Though we do not have the power to control the real effects of our own actions, politics rallies its public on the assumption that something is at stake, and what is at stake is how things will be tomorrow as a result of our actions today. The political understanding of the design process thus leads to a 'critique' in the Kantian sense of the term. This critique appreciates the designer fallacy with its technological hubris as a heuristically valuable fiction that calls for a proactive ethics of correct wishing. At the same time this necessary fiction is exposed as a dangerous conceit that demands an ethical corrective, namely a humble appreciation of contingency and human limits of knowledge and control.

It is in terms of this fictional *as if* that one can finally frame responsible research as the design of 'moralising things'. Verbeek follows Latour and Achterhuis by construing these in terms of delegation: the human designer inscribes into devices a script such that interactions with the device regulate human behaviour according to some moral precept—most famously, perhaps, the 'Berlin key', which opens a door only for those who diligently lock it again afterwards (Latour 1991). One might also adopt a rather more virtue–ethical approach according to which artefacts do not take over human responsibilities, but are themselves virtuous. Here, then, responsible design aims for devices that are frugal rather than wasteful, reliable rather than temperamental, unassuming rather than intrusive, transparent rather than opaque, friendly rather than difficult to operate, or safe rather than risky. It is the designer fallacy that leads us to believe that we can create a better world by designing moralising or virtuous things, and as such design projects do and ought to involve society at large, they extend far beyond the laboratory floor. Responsible research will inform and participate in the political struggle for a better world, especially where its design requires scientific and technological research, but, at the same time, it will question any exaggerated faith in the power of technology to create a better world by way of controlling or designing its own effects and the many ways it mediates human agency.

In summary, then, the 'ethics of design' perspective points in the right direction, literally. It points away from the deep dark crevices of the laboratory where the nature of decisions lies beyond the competence of citizens and where the scope of decision is so restricted that only through artificial inflation by constructivist accounts these decisions can appear to cumulatively set the course for technological development. Instead of buying into this picture of the laboratory as the breeding ground for the future, the 'ethics of design' ties the work of researchers and developers back into a larger societal process—a superficial sign of this 'reconnect' is the difficulty, if not impossibility, even, to designate the researchers and developers and to circumscribe a specific site at which collective design processes take place. As the design begins with the identification of needs, with funding or product development decisions, and as it ends with the establishment of patterns of use by clients or consumers, research in a design mode holds no place of privilege but contributes, among many other activities, to an overarching dynamics.

From the point of view of the laboratory floor, the responsibility of researchers would require a stepping-back and a refusal to accept a brief that overestimates what goes on in the laboratory as a distinct site. Instead, responsible research should be conceived as civic involvement in the overall project of shaping the world, that is of identifying the problems that need solving, deliberating potential solutions, and, if these solutions are to involve scientific or technological research, to submit specific design choices to processes of public reasoning. In and of itself this reorientation effects what I have called a shift from ethics to politics of research (see note 1).

Politics of research

Though an 'ethics of design' points in the right direction, namely the politics of a kind of research that serves societal projects to change the world in this way or that, it is confronted with two difficulties. First, it requires that researchers and science policy-makers, as well as philosophers, stop privileging scientific research and the laboratory as the site where the future is made. One therefore has to give up conceptions of science that are very familiar and very dear to us in order to understand contemporary research as a technoscientific enterprise that is undertaken in a design mode. The second difficulty is more severe in that by the very notion of 'design' we may lose again what was gained in the shift to 'design ethics'. As we can learn from Verbeek and Kiran's sophisticated reflections, the notion of design leads into the designer fallacy—and attempting to escape the fallacy is to lose the meaning of 'design'. Instead of acknowledging that responsible research contributes to a process of making the world in a haphazard, well-intentioned, evolutionary manner with ultimately unpredictable results, 'design ethics' suggests that our technological future is an object of design and that it can be anticipated, evaluated and modulated because, after all, it might turn out to be just what we are now envisioning it to be. Instead of serving as a necessary corrective to a regime of technoscientific promising (Felt et al. 2007), of control fantasies, of technological hubris or just plain hype, 'design ethics' threatens to further amplify exaggerated expectations of technological solutions to today's global or societal problems.

Note

1. As I am employing these terms here, ethics and politics are not distinguished by the issues to be deliberated ('Should we use biocompatible materials?' or 'How can we take the actual needs and preferences of patients into account?'). Instead, they are distinguished by the way in which these issues are deliberated and finally settled. I refer to ethics if this is a determination in respect to an opinion, principle, attitude or action of what is to be considered good, right or appropriate. I refer to politics where one opts for public participation in an open-ended and contested process that amalgamates economic interests, religious dogma, environmental concerns, personal values, human rights and much more. In the case of ethics, responsible researchers would seek at least temporary closure (long enough at least to justify the next action taken or the next opinion to be pronounced). Responsible research contributes to the sphere of politics by making issues amenable to public decision-making and debate.

References

Böschen, S. and Wehling, P. (2004) *Wissenschaft zwischen Folgenverantwortung und Nichtwissen: Aktuelle Perspektiven der Wissenschaftsforschung* (Wiesbaden: VS Verlag für Sozialwissenschaften).
Davies, S., Macnaghten, P., and Kearnes, M. (2009) *Reconfiguring Responsibility: Lessons for Public Policy*, Part 1 of the Report on Deepening Debate on Nanotechnology (Durham: Durham University).

European Commission (2008) *Conduct for Responsible Nanosciences and Nanotechnologies Research* (Brussels: European Commission).

Felt, U., Wynne, B., Callon, M., et al. (2007) 'Taking European Knowledge Society Seriously', Report of the Expert Group on Science and Governance to the Science, Economy and Society Directorate, Directorate-General for Research, European Commission (Brussels: European Communities).

Fisher, E. and Mahajan, R. (2006) 'Midstream Modulation of Nanotechnology in an Academic Research Laboratory', in *Proceedings of IMECE2006: American Society of Mechanical Engineers International Mechanical Engineering Congress and Exposition*, 5–10 November, Chicago, IL, USA.

Gamm, G. (2000) 'Technik als Medium. Grundlinien einer Philosophie der Technik', in Gamm, G. (ed.) *Nicht nichts. Studien zu einer Semantik des Unbestimmten* (Frankfurt: Suhrkamp).

Heidegger, M. (1993) 'The Question Concerning Technology', in Krell, D. (ed.) *Basic Writings* (New York, HarperCollins).

Ihde, D. (2008) 'The Designer Fallacy and Technological Imagination', in: Vermaas, P.E., Kroes, P., Light, A., and Moore, S.E. (eds) *Philosophy and Design. From Engineering to Architecture* (pp. 51–9) (Dordrecht, Springer).

Kiran, A. (2012) 'Responsible Design. A Conceptual Look at Interdependent Design–Use Dynamics', *Philosophy and Technology*, 25: 179–98.

Latour, B. (1990) 'The Force and Reason of Experiment', in Le Grand, H. (ed.) *Experimental Inquiries, Historical, Philosophical and Social Studies of Experimentation in Science* (pp. 48–79) (Dordrecht: Kluwer).

Latour, B. (1991) 'The Berlin Key or How to Do Things With Words', in Graves-Brown, P. M. (ed.) *Matter, Materiality and Modern Culture* (pp. 10–21) (London, Routledge).

Nordmann, A. (2007) 'Knots and Strands: An Argument for Productive Disillusionment', *Journal of Medicine and Philosophy*, 32(3): 217–36.

Nordmann, A. (2010) 'A Forensics of Wishing: Technology Assessment in the Age of Technoscience', in *Poiesis & Praxis*, vol. 7 (pp. 5–15) (Dordrecht: Springer).

Nordmann, A. and Schwarz, A. (2010) 'Lure of the "Yes": The Seductive Power of Technoscience', in Kaiser, M., Kurath, M., Maasen, S., and Rehmann-Sutter, C. (eds) *Governing Future Technologies: Nanotechnology and the Rise of an Assessment Regime* (pp. 255–77) (Dordrecht: Springer).

Nordmann, A., Radder, H., and Schiemann, G. (eds) (2011). *Science Transformed? Debating Claims of an Epochal Break* (Pittsburgh: Pittsburgh University Press).

Nowotny, H., Scott, P., and Gibbons, M. (2003) ' "Mode 2" revisited', *Minerva*, 41: 179–94.

Schwarz, A. and Nordmann, A. (2010) 'The Political Economy of Technoscience', in Carrier, M. and Nordmann, A. (eds) *Science in the Context of Application* (pp. 317–36) (Dordrecht: Springer).

Sokal, A. (1996) 'Transgressing the Boundaries: Towards a Transformative Hermeneutics of Quantum Gravity', *Social Text*, 46/47: 217–52.

Verbeek, P.P. (2011) *Moralizing Technology; Understanding and Designing the Morality of Things* (Chicago/London: University of Chicago Press).

Vogt, T. (2010) 'Buying Time – Using Nanotechnologies and Other Emerging Technologies for a Sustainable Future', in Fiedeler, U., Coenen, C., Davies, S., and Ferrari, A. (eds) *Understanding Nanotechnology: Philosophy, Policy and Publics* (pp. 43–60) (Heidelberg, Akademische Verlagsgesellschaft AKA).

von Schomberg, R. (2010) 'Organising Collective Responsibility: Ion Precaution, Codes of Conduct and understanding Public Debate', in Fiedeler, U., Coenen, C., Davies, S., and Ferrari A. (eds) *Understanding Nanotechnology: Philosophy, Policy and Publics* (pp. 61–70) (Heidelberg: Akademische Verlagsgesellschaft AKA).

Winner, L. (1993) 'Upon Opening the Black Box and Finding It Empty: Social Constructivism and the Philosophy of Technology', *Science, Technology and Human Values*, 18(3): 362–78.

Index

Printed in the United States
by Baker & Taylor Publisher Services